THEMES IN GEOMORPHOLOGY

Themes in Geomorphology

Edited by Alistair Pitty

CROOM HELM
London & Sydney

Croom Helm Ltd, Provident House, Burrell Row,
Beckenham, Kent BR3 1AT
Croom Helm Australia Pty Ltd, First Floor, 139 King Street,
Sydney, NSW 2001, Australia

British Library Cataloguing in Publication Data
Themes in geomorphology.
1. Geomorphology
I. Pitty, Alistair F.
551.4 GB401.5
ISBN 0-7099-2066-0

Printed and bound in Great Britain

CONTENTS

INTRODUCTION

This book conveys the flavours of geomorphology and the scents of its ideas. It offers a geomorphological aperitif for students of geography to savour whilst contemplating choices for a main course. Themes in geomorphology are illustrated from the experience of the contributors who are well-known for their persistent attention to a specific aspect or regional setting of landform study.

There are three main settings which colour geomorphological conjectures and become both a strength and weakness in landform study. First, it is well known that, regionally, geomorphology was initially forged by the likes of German geographers in the Saharan heat or by the pioneers of geological surveys, amid the landscape spectacles of the arid southwest of the United States. In addition to the distinctiveness of climatic zones, as described by the research and reviews of French geomorphologists such as A. Cailleux and J. Tricart, a second group of themes emerges from distinctive geological settings. For example, C.A. Cotton was able to link Davisian geomorphology of the 1900s with advancing thought a generation later, greatly aided by the rich array of landforms present in a tectonically active zone like New Zealand. Thirdly, perceptions are refocused at particular junctures in time, as well as reflecting their spatial settings. Thus, 'new' themes periodically emerge, partly due to the circumstances of a particular time. For instance, a deliberate drive by the US Office of Naval Research accelerated the first notions of modern geomorphology in the late 1940s.

Conversely, some of the most familiar 'themes' in geomorphology have emerged as perceptions, peculiar to a distinctive setting and were ambitiously extended more broadly. The classic examples include, regionally, debates stemming from L.C. King's suggestions that the pediment forms of semi-arid southern Africa are recognisable in other climatic zones. In terms of geological environment, W.M. Davis recognised that his simplifying postulate of periodic 'stillstands' of earth structures did not apply in tectonically active zones. He therefore advised a re-learning of one's geomorphological A-to-Z on moving into an area like California. As in all branches of advancing knowledge, geomorphologists are not always critical when embracing new themes nor sufficiently unsentimental in discarding the outdated.

Nonetheless, it is hoped that the present book portrays the positive features of pluralism in geomorphology rather than its frictions. It is structured around these three main sources from which themes in geomorphology stem. In its larger part, the first nine chapters focus on processes operative, and their associated landforms in distinctive climatic zones, including coastal margins in Chapter 4. Three chapters then feature the distinctive geological settings of karst, volcanicity, and tectonic activity. Finally, three chapters introduce emerging themes of our time, in which historical photographs are now sufficiently dated to be compared with changed landforms of the present-day, urbanisation advances, and technology reaches out towards the stars.

Acknowledgements

The publishers, contributors, and editor would like to add the following acknowledgements to those included within the text. We are grateful to the Editor of *Geografiska Annaler A* for Figures 3.3 (and L. Vilborg), 3.4 (and M. Clark), 3.6 and 3.7 (and H. Svensson); Blackwell's Scientific Publications Ltd, for Figure 5.2; Hunting Surveys Ltd for Figure 9.2 and the Editor of *Lethaia* for Figure 9.10; Dr A.C. Waltham for lively exchanges about lost caverns and the enclosed depressions discussed in Chapter 10; the New Zealand Department of Lands and Survey for Figure 11.1; the United States National Science Foundation for the financial support of Grant EAR 8119932 during the development of the technique described in Chapter 13 and Herbert Verville for the final draughting of its diagrams; Edward Arnold Publishers, for Figure 14.6, reproduced from Ian Douglas's *The Urban Environment.*

In addition, I am grateful to all authors for their valuable contributions, genial co-operation, and for the many, subtle, and perceptive ways in which they have lent general encouragement and support. I have benefitted from exchanges with many other geomorphologists whilst the 'Themes' have taken shape, and distinctive features of the collection stem from the counsels of Professor J.H. Appleton. Editorial attention to details of the texts and bibliographies was facilitated by access to the resources of the Brynmor Jones Library, University of Hull.
Alistair F. Pitty
Barton-on-Humber

1 ASPECTS OF GLACIERS AND THEIR BEDS

W.H. Theakstone

Introduction

A knowledge of the processes which fashion landforms is necessary for understanding landscapes. Therefore, studies of the incorporation, transport and deposition of materials by glaciers should combine direct observation of the processes which operate beneath glaciers with sound glaciological theory. As opportunities for observation are limited, many discussions of sub-glacial conditions have been based on simple assumptions, involving two-dimensional models, beds of solid rock and pure, isotropic ice. In recent years, the development of hydroelectric power schemes taking water from beneath glaciers has increased opportunities for subglacial studies, and some glaciological theories are now more realistically based.

If a glacier has been moving over an area for a very long time, erosion will probably have removed much of the sediment and weathered debris which may have been present in the past. Areas of bare rock are common around the margins of retreating glaciers, and most observations beneath existing glaciers have been made at sites where the bed is solid. However, when the bottoms of boreholes made between 1969 and 1976 through the Blue Glacier, Washington, USA (Figure 1.1. A) were examined with the aid of small television cameras, the bed invariably was found to consist of rock debris. The glaciologists, having noted the abundance of abraded and striated rock at the glacier margin, were surprised at their failure to observe any solid rock at the bottom; the basal conditions did not correspond closely to those assumed in existing theories of glacier sliding.

Advancing glacier fronts generally move over areas with a sometimes substantial, cover of soil, sediment or weathered rock. Sub-glacial deformation of this material may account for much of the total basal movement of a glacier. Theories of glacier sliding involving such deformable bed material are much less common than are theories relating to a solid bed.

Figure 1.1: Four glaciers at which sub-glacial and glacio-hydrological studies have been made. A. Blue Glacier, Washington, USA. B. Glacier d'Argentière, Mont Blanc, France. C. Bondhusbreen, Folgefonni, Norway. D. Austre Okstindbreen, Okstinden, Norway. The glacier beds have been investigated in natural cavities beneath the ice, in tunnels made to collect water and remove sediment for hydroelectric power schemes, and through boreholes drilled from the glacier surface. Note that two rivers issue from Austre Okstindbreen, the northern one flowing back under the glacier after a short subaerial course.

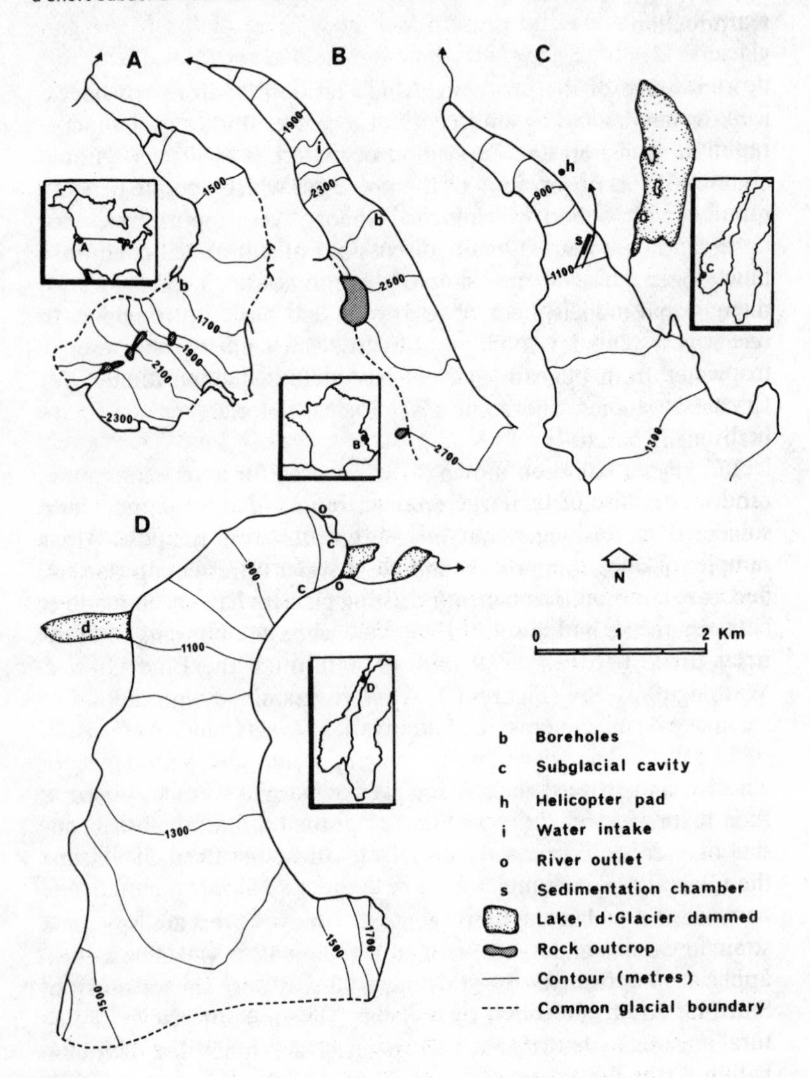

The nature and behaviour of ice at the bottom of glaciers

The bottom of a glacier may be thermally complex; even if the basal ice generally is temperate (i.e. at the melting-point), patches of cold ice with below-melting-point temperatures may occur. The basal ice may include both true glacier ice (metamorphosed snow) and refrozen meltwater; some may be bubble-free; elsewhere, many bubbles may be present. At the bottom of the Norwegian glaciers Østerdalsisen and Austre Okstindbreen, cavities form down-glacier of steeply inclined parts of the bed, and some basal ice is trapped between the rock floor and ice which is moving more rapidly over it. This forms thin zones, characterised by numerous air bubbles and low bulk density. The bubble-rich ice deforms rapidly (Theakstone, 1979).

The debris content of ice adjacent to a glacier bed may be as much as 30 per cent by volume. However, the proportions of relatively clean and dirty ice in contact with the bed must vary from one part of a glacier to another. Most basal ice is markedly anisotropic and frequently displays well-developed banding or foliation. Crystal sizes differ: the high shear induced by irregularities of the bed may result in the development of zones of very fine-grained ice. Crystal orientation patterns (ice fabrics) are likely to be non-random because of the large total strain to which the ice has been subjected in its long history. If sufficient stress is applied to a sample of ice, the rate at which it deforms (the strain rate) becomes constant. Laboratory studies have shown that the relation between the deformation of samples of polycrystalline ice and the stress producing the deformation has the form:

$$\dot{\varepsilon} = A\tau^n$$

where $\dot{\varepsilon}$ is the strain rate, τ is the applied stress, n is a constant and A is a function of the pressure, temperature, texture (shape and size of individual crystals) and fabric. Although this relationship, the Glen flow law, is much used in theories of glacier sliding, comparison of the behaviour of the glaciers and of samples of ice strained in laboratory investigations indicates that unmodified application of the law to glaciers is not justified (Hooke, 1981). Since ice strain is affected by its temperature, grain size and structural anisotropy, and by the sediment particles and other inclusions within it, the behaviour of ice close to the bed of a glacier should

Figure 1.2: Folded basal ice at the Norwegian glacier Corneliussens Bre. This sediment-rich basal layer contrasts with the overlying, relatively clean ice. Boulders within the cavity beneath the glacier are mud-coated. The tape holder in the centre is 5 cm across.

not be expected to conform to a simple flow law. As ice deforms, the textural and structural changes which occur may influence its subsequent behaviour significantly. The flow law at any part of a glacier therefore must be a function of location, defined by the local crystallographic fabric and stress situation. The flow law derived from laboratory studies of isotropic ice is so modified by local factors that it may underestimate actual strain rates by a factor of more than 10.

The deformation of dirty ice is affected by the concentration of the debris particles. Although low concentrations of sediment within the ice may increase the strain (creep) rate, concentrations in excess of about 10 per cent by volume apparently stiffen ice, with the particles impeding creep. Very dirty ice, like ice-rich soils, has a creep rate much lower than that of clean polycrystalline ice. Clearly, the mechanical properties of ice at the bottom of a glacier cannot be characterised without specifying such other properties as impurity or inclusion content, texture and fabric. Ice passing round the sides of upstanding parts of a glacier bed, or round other obstacles such as boulders, is subject to transverse compression and longitudinal extension. Upstream of the obstacle, longitudinal compression occurs, whilst longitudinal extension is characteristic of ice moving over the top of the obstacle. Such deformation patterns may lead to folding of basal ice and may cause sediment to move towards cavities and low pressure zones in the lee of obstacles. Strong basal folding of basal ice can cause a single debris-rich layer to appear several times in vertical section, thereby increasing the overall concentration of sediment in ice close to the bed (Figure 1.2).

Pressure fluctuations at glacier beds change ice melting-point, and water may be squeezed out of the ice by high pressure. The water tends to flow towards zones of lower pressure, such as cavities at the bed. Some refreezes there, forming a basal layer of bubble-poor ice with dirty layers containing dispersed sediment particles and rock fragments. The chemical composition of such regelation ice frequently differs significantly from that of glacier ice; and the impurities within it may impede sliding. Formation of a patch of cold basal ice, as water produced by pressure melting within the ice is squeezed out, may eliminate a thin film of water which otherwise would be present (Robin, 1976). Such cold patches at the bed must influence glacier sliding. The ceilings of natural sub-glacial cavities may be below melting-point at atmospheric pressure, reflecting the lower melting-point of ice under stress up-glacier of the cavity. Some water expelled from the glacier refreezes at the cavity ceiling when the pressure is released, forming ice stalactites and excrescences (Figure 1.3) or rows of small bumps. These may be related to variations of sliding of the ice up-glacier of the cavity (Andreasen, 1983).

Glacier sliding

Theories

Glacier flow is commonly described in terms of two components – plastic deformation within the glacier, and sliding of the glacier over its bed; the latter usually occurs only where the basal ice is temperate. The relationship between the sliding velocity of a glacier, the characteristics of its bed and the basal shear stress may be expressed in terms of a sliding law. However, the nature of glacier sliding remains relatively poorly understood; it may involve intense deformation of a layer of ice close to the bed, rather than

Figure 1.3: Ice accretions on the wall of a natural cavity beneath the Norwegian glacier Austre Okstindbreen. Ice flow is from right to left. The near-parallel flow lines along the wall result from moulding of the ice as it flows past bedrock irregularities up-glacier of the cavity. Platy ice structures, projecting from the wall, probably develop as water which has been extruded from the glacier by ambient hydrostatic pressure refreezes. The thin ice stalactite in the centre is about 15 cm long.

simply discrete motion of the body of ice over the underlying surface.

As glacier beds are not smooth, explanations of glacier sliding have to account for the movement of ice past irregularities of the bed. In 1957, Weertman suggested two possible mechanisms: enhanced plastic deformation, and regelation. Although all the ice of a glacier is likely to be deforming plastically, the higher stress against the up-glacier side of obstacles at the bed causes above-average deformation. Regelation involves melting of temperate ice as a result of the higher pressure present at the up-glacier side of obstacles; the meltwater refreezes at the down-glacier side, where the pressure is lower. The resistance to sliding offered by the irregularities of the bed (i.e. its roughness) varies with their size. Enhanced plastic deformation permits ice to overcome the resistance of large obstacles, whilst regelation allows ice to flow easily round small ones. Weertman suggested that there was a controlling obstacle size at which overall resistance to the two mechanisms was greatest and that this critical size largely determined a glacier's sliding velocity. Some recent experimental work has indicated that sliding may be controlled more by ice deformation than by regelation and that the regelation phenomenon itself may be the result of deformation.

Weertman's early theory of glacier sliding has been modified to take account of several additional factors. These include the separation of parts of the glacier bottom from the bed, with the formation of cavities on the down-glacier side of obstacles; these reduce the friction between ice and rock. Short-term variations of glacier surface velocity, assumed to result from changes of sliding velocity, have also been considered. An increase of thickness of the water layer present at the glacier bed may result in the submergence of some of the smaller obstacles which previously resisted sliding, thereby reducing the effective roughness of the bed.

Although the behaviour of water at the bed is an integral part of the process of glacier sliding, it has not yet been incorporated successfully into sliding theories. Some water must be present at any point at which temperate ice is sliding over a rock bed. Additional water is formed by frictional heating and by geothermal heat flow from the bed to the basal ice. Water formed by melting at the glacier's upper surface is more abundant, and much reaches parts of the bed. If the glacier bed is impermeable, as in most theoretical sliding models, the water must flow along it, thereby affecting both the effective roughness and the sliding velocity. However, if the bed is permeable, water may move through it to the glacier bottom, or water may move from the glacier into the bed. Many observations have shown that glacier surface flow rates and, by implication, glacier sliding velocities, vary with the availability of water: glaciers generally move more rapidly in the summer melt season than in winter. However the relationship between velocity and water availability is not simple: the pressure of the sub-glacial water is a significant factor.

Observations

Direct observations of sliding have been made in tunnels excavated to glacier beds and in natural sub-glacial cavities accessible from the glacier margins. Indirect determination of sliding rates utilizes boreholes drilled from the glacier surface to the bed (Figure 1.1 A).Movement of the top of the borehole and the inclination and bearing (azimuth) of the borehole axis at different depths are measured at different times, commonly about one year apart. Most glacier boreholes have been made with drills which melt their way down through the glacier. Such thermal drills may fail to penetrate englacial debris and, more generally, basal layers of dirty ice. Photography from the bottoms of boreholes which had been extended with the aid of electromechanical drills and removal of rock debris indicated that sliding of Blue Glacier was much less than that calculated indirectly from changes of borehole position and inclination. Previous assessments of glacier sliding made in this way were called seriously into question (Engelhardt *et al.*, 1978).

Frequently, borehole inclination data, showing the variation of velocity with depth below the glacier surface, have been plotted two-dimensionally, with the implication that the direction of flow is uniform. However, measurements of an array of boreholes at Athabasca Glacier, Canada indicated that the flow direction varied with depth, and the photographic studies from the Blue Glacier boreholes revealed marked differences of direction of surface and basal flow. Although variability of flow direction within a glacier has been given relatively little attention in theoretical studies and in glacier flow modelling, both direct observations and the evidence provided by striations on bedrock indicate that it is an important characteristic of movement of ice at the bed. Folding of ice which is subjected to compression as it comes into contact with the bed at the downstream limits of subglacial cavities is common (Theakstone, 1966). The strong lateral flow component of the deforming ice reflects the sometimes complex relationship of glacier flow direction and the three-dimensional geometry of the bed.

At many glaciers, debris is present in the lowermost few centimetres of ice. Some particles partly within the ice make contact with the bed where they are retarded relative to the moving ice, because of the frictional drag, and particles entirely surrounded by ice may tend to form folds over them. When debris concentrations are high and drag is large, the basal debris determines the sliding velocity. Frequent particle collisions result in accumulations forming, and ice is forced to move round such mobile obstructions as well as round the fixed obstacles of the rock bed (Boulton *et al.*, 1979). As seen below the Glacier d'Argentière in the French Alps (Figure 1.1. B), the resultant transverse component of flow of basal ice causes debris-rich ice and particle concentrations to move through furrows alongside the obstructions. Water at the bed is often diverted into such furrows, so that more ice melts there than above the adjacent hump, increasing still further the differential concentration of debris at the bed and its effect on glacier sliding and erosion.

Direct measurements of sliding rates beneath several glaciers including Blue Glacier, the Glacier d'Argentière, Østerdalsisen and Austre Okstindbreen (Figure 1.1. C), have shown that marked variations occur. Close to the bed, motion of the ice has a stick–slip nature. This may result from spatial and temporal variations of basal pressure causing ice to freeze to the rock bed. Such a frozen patch experiences an ever-increasing stress as the glacier above the basal layer moves forwards, until formation of a crack relieves the stress. Should stresses and velocity changes near the transition from a cold, non-sliding zone to one of sliding be sufficiently large, rock may be removed from the bed, contributing to the accumulation of debris, such as that seen at the bottom of Blue Glacier boreholes. Stick–slip motion is the most probable explanation of an apparent annual cycle of seismic events recorded in the vicinity of glaciers in the Cascade Range, Washington, USA; activity peaks in April–September, and is at a minimum in December–February. Under the Glacier d'Argentière, strain events recorded in 1975–76 were twice as frequent in September as in December–January, and nearly three times as frequent as in April, when sliding was slowest. The seasonal pattern may be explained in terms of changes of basal water film thickness, water pressure, thermal state and regelation heat flow.

Glacier hydrology

The nature of the hydrological systems

Since the availability of water influences the rate of glacier sliding, information about the nature of water movement at glacier beds is of great value to glaciologists. As rivers transport most of the sediment leaving a glacier — about 90 per cent of that from the Norwegian glacier, Bondhusbreen (Figure 1.1. D), for instance (Hagen *et al.*, 1983) — they must have a significant role in processes active at the bed.

Problems of access limit direct observations of sub-glacial hydrological systems. Different theories assign dominant roles to thin films of water at the bed, conduits incised into the bed, conduits largely within basal ice and water-filled cavities. Sheetflow of water beneath a temperate glacier tends to be unstable as differences of thickness tend to increase. Much water is stored in cavities at the bed, which are areas of relatively low water pressure. Together with the water flowing through the conduit systems, this storage could account for a significant part of the total annual production of meltwater. Much winter precipitation is stored until the following summer, when it contributes to the considerably larger flow of glacier rivers. Eighty-five per cent of the total annual river discharge from Bondhusbreen occurs between June and September. For part of the summer, discharge exceeds the input of water supplied to most glaciers by surface ablation. Diurnal variations of flow are characteristic during rainless periods in summer: maximum discharge generally occurs a few hours after peak melting, and there is a minimum at night. The fluctuations are superimposed on a baseflow component which declines if surface melting is interrupted by a prolonged cold period or a fall of snow.

At most valley glaciers, water emerges in a single river or in no more than two or three separate courses — at Bondhusbreen, for example, 90 per cent of the summer discharge and all the winter discharge is accommodated in a single course. Some glacier rivers have relatively stable courses: that beneath Bondhusbreen has occupied the same course since it was first collected sub-glacially for hydroelectric power purposes in 1978. Elsewhere, however, shifts may occur. For several years, the main river beneath the Glacier d'Argentière occupied a channel incised into the central part of the rock bed, but it disappeared in the spring of 1976. Three years later, the river was found near one side of the glacier and has remained there subsequently.

At glaciers which supply two rivers, changes of position of the major outflow from one year to another have been noted. The changes generally occur between summers; once one outlet is established as the major one, a change during the summer is unlikely. Two rivers issue from Austre Okstindbreen. In five of the summers between 1977 and 1983, the one close to the left side of the terminus was the major outflow, and the discharge of the river close to the right side was low. In the other two summers, however, most of the water left the glacier in the river at its right side.

A theoretical model of glacier hydrological systems developed by Röthlisberger accounts for water flow, not only in a basal conduit along the deepest part of a glacier-covered valley floor, but also in lateral streams at the level to which water would rise in boreholes which penetrated to the glacier bed (the piezometric level or hydraulic grade line). In a recent theory developed by Lliboutry, however, sub-marginal streams (gradient conduits) are likely to survive only alongside over-deepened parts of the valley; elsewhere, the pressure gradient drives the water to the lowest part of the valley floor.

Conduits beneath and within glaciers tend to close as a result of the pressure exerted by the overlying ice. This is opposed by frictional heating: water flowing through the conduits tends to melt the walls. As the ice pressure cannot vary rapidly at a point, short-term fluctuations of water inputs may determine the degree to which conduits are water-filled. The water may be at atmospheric pressure for most of the time, but the conduits may be filled completely during periods of peak discharge, and so experience relatively high water pressures. With the decline of melting at the end of the summer, conduit size decreases; parts of the system which contain no water close down completely during the winter. High water pressures noted beneath some glaciers in winter probably reflect the small size of the conduits and the limited amount of melting of their walls. In early summer, water inputs to glacier hydrological systems increase rapidly and, because the relatively small passages become filled, the water pressure within conduits again is high. It is at this time, as conduit systems are developing, that between-summer changes, such as those noted at Austre Okstindbreen, are likely to occur. However, as the summer progresses, conduits widen and, despite the higher discharge, the mean water pressure falls again. Water pressure fluctuations in sub-glacial conduits may be responsible for erosion and sediment incorporation. If the local water pressure falls to the level of its vapour pressure, air pockets form near the solid walls. When the pressure rises, the pockets break up, causing micro-jets to form in the direction of low pressure. In striking the conduit wall, these jets create a very strong local pressure surge. Repetition of this pressure of cavitation causes considerable erosion.

Elements of the geometry of the sub-glacial drainage network may be retained on rock surfaces, to be exposed by subsequent glacier retreat. Observations in front of two small Canadian glaciers suggest that a thin sheet of water over most of the bed coexisted with interconnecting channels and cavities which drained most of the meltwater, but that it was relatively independent of the channels. Calcium carbonate precipitates coating some channels indicate that the sub-glacial hydrological system changes with time.

Borehole observations

There have been few successful attempts to monitor sub-glacial water pressures directly. Boreholes drilled through several glaciers indicate that basal hydrological systems are highly variable, both in space and time. Of 13 boreholes made at South Cascade Glacier, Washington, USA, in 1973 and 1974, 5 apparently reached parts of an active basal system, but only 2 of the 24 holes drilled in 1975 and 1977 made such a direct contact, suggesting that most of the glacier bed is hydraulically inactive and isolated from a few active sub-glacial conduits. In some boreholes, both at South Cascade Glacier and at Blue Glacier, the water level remained high and almost constant, suggesting that little leakage occurred from the bottom. High water levels may also indicate that a borehole is connected to a sub-glacial conduit in which the pressure is high, or one which simply cannot accommodate all the available water. If the pressure of the water on the bed at the bottom of a borehole is higher than that of the surrounding, less dense ice, leaks will develop, probably spreading gradually away from the borehole until connecting with an active sub-glacial conduit system. A few boreholes have emptied on reaching the glacier bed and have remained dry, suggesting that a connection has been made with a sub-glacial drainage system at atmospheric pressure. Some fluctuations of borehole water level result from the inability of conduits to respond rapidly to changes of supply. Only when synchronous fluctuations occur in several boreholes can it be concluded that they are linked to a drainage system below the glacier. In such cases, the differences of level (piezometric gradient), indicating the head loss between boreholes, may be a valuable clue to the sub-glacial water flow direction. Thus, in June 1980, five boreholes along a line across the Glacier d'Argentière overlying an over-deepened part of the valley floor apparently were linked together and to a sub-glacial conduit near the right side of the glacier which controlled the water pressure; the piezometric gradient was towards the valley side. About 170 metres away, however, up-glacier of the over-deepened zone, the piezometric gradient in August 1980 was directed towards the central, deepest part of the valley (Hantz and Lliboutry, 1983).

Tracers, isotopes and electrical conductivity

Information about the form of the hydrological systems linking points of inflow of water at a glacier's surface and points of outflow generally is scarce. Tests in which dye or salt is injected into streams entering moulins indicate that some water passes very rapidly through a glacier at velocities equivalent to those of open channel flow elsewhere, and that some is delayed in transit for a few hours or days. High through-flow velocities indicate that efficient interconnected systems link supraglacial stream channels, moulins and some crevasses with the main glacier river courses. Velocities within the range 0.4–1.2 ms^{-1} have been recorded at several glaciers; at Austre Okstindbreen, water from some moulins flows at mean velocities as high as 1.8 ms^{-1} over distances in excess of 1.4 km. Tests which fail to yield positive results suggest that the water into which tracer was injected at the surface went into longer-term storage in or below the glacier. Some water may be stored for several weeks or months. At Austre Okstindbreen and at some Swedish glaciers, tracer tests have shown that discrete internal drainage systems are related to surface crevasse patterns which tend to direct water towards the margins of the glaciers. Links between the systems may exist at certain times during the summer. At Pasterzengletscher, Austria, water entering some moulins sometimes emerges in both rivers flowing from the glacier; at other times, the same moulins are connected to only one of the outflowing streams.

Analyses of the water leaving a glacier may provide some information about the hydrological systems through which it has passed. In addition to meltwater formed from snow and from ice, outflows may include some groundwater and, at times, rainwater. Isotopic analysis has provided much information about sources and their variations with time. Part of the water has passed relatively rapidly through the system, and part has been delayed. Since the latter is likely to have been in contact with sediments and weathered rock debris, variations of the dissolved load may reflect changes of proportion of the delayed component.

The isotopic composition of the first rain to fall from an air mass is similar to that of the ocean water from which the water vapour was derived. However, because heavier water molecules condense more rapidly from vapour than do lighter ones, the ratios of the concentrations of heavy and light atoms of hydrogen (D/H) and of oxygen ($^{18}O/^{16}O$) in precipitation change with time, with distance from the sea and with altitude. The isotopic ratios in precipitation depend on the temperature: the lower the temperature, the more efficient is the separation of the isotopes. Winter precipitation has a lower $^{18}O/^{16}O$ ratio than that which falls in summer. In addition to this seasonal effect, the composition of falling rain and snow varies on other timescales — during and between meteorological events and over longer periods. Glacier meltwater generally has a lower $^{18}O/^{16}O$ ratio than has summer precipitation. As sub-glacial aquifers which are depleted in summer and recharged in winter have a cold isotopic signature, groundwater contributing to glacier river discharge in summer tends to have a relatively low $^{18}O/^{16}O$ ratio.

During the late 1960s and early 1970s, measurements in the Austrian Alps revealed that concentrations of isotopes in samples of snow, ice and riverwater could be used to determine the relative contributions of meltwater from snow, meltwater from ice and sub-glacial spring water to the discharge of a glacier river during summer. At Hintereisferner, spring water accounted for 40 per cent of the run-off in the months of July to September. Through the summer of 1980, the isotopic composition of water issuing from Austre Okstindbreen in early afternoon showed a general trend towards lower $^{18}O/^{16}O$ ratios. Apparently, the contribution of snow to the discharge of the glacier river was decreasing. In the summer of 1982, samples of water from the main glacier river were isotopically heavier than those collected at the same time from a small stream which issued at the other side of the glacier; the stream was probably supplied principally by snowmelt. Early in the melt season, the discharge of the main river decreased throughout a five-day period which followed a spell of persistent rain, but the variations of isotopic composition of the water were slight, suggesting that much of it may have been displaced from storage. Diurnal variations of isotopic composition in phase with variations of river discharge have been noted during periods of fine weather at several glaciers, but such patterns disappear in wet periods. Analyses of precipitation and of glacier riverwater during a storm at the Neves Glacier, Italy, indicated that, during the rising limb of the hydrograph, discharge consisted mainly of pre-storm meltwater displaced from the glacier. The peak of the rainfall component was delayed, arriving after the peak of the displaced meltwater component (Van de Griend and Arwert, 1983).

The information supplied by isotope analysis of water samples may be supplemented by studies of their ionic composition. The more ions present, the higher is the electrical conductivity of the water. Glacier meltwater is a very poor conductor but, as the total dissolved solids concentration increases, so does the electrical conductivity. Solutes may be added to water which takes a sub-glacial route and encounters sediments at the bed or in basal ice. Enrichment occurs when slow-moving meltwater first encounters sediments. Continuous monitoring of electrical conductivity at the river flowing from the Gornergletscher, Switzerland indicated a pattern of night-time maxima and daytime minima. The out-of-phase relationship between river discharge and the content of dissolved solids suggested that much of the discharge during daily peaks was of water formed by melting at the glacier surface which left without having undergone chemical enrichment.

As at many glaciers, the principal drainage system at Gornergletscher probably consists largely of ice-walled conduits through which water passes quickly from the surface to the terminus. As the meltwater supply to the system increases during the day and parts of the conduits become filled, the water pressure rises. Should it exceed the hydrostatic pressure of the adjacent ice, leakage occurs. Cavities form as ice is separated from the bed. As long as water is forced into them under pressure, they continue to grow. Later in the day, as surface melting decreases, the chemically enriched water returns to the conduits. Collins (1979) suggested that about 30 per cent of the water flowing within Gornergletscher spends at least 12 hours at the base. Although the conductivity of water flowing from Gornergletscher is much higher in winter than in summer, discharge is very low; almost all the solute load from beneath the glacier leaves during the summer when high flows flush out solutes (Collins, 1981). The small amount of water flowing from the glacier in winter probably has had substantial contact with basal sources of solutes. Elsewhere, sub-glacial groundwater may contribute to high conductivity values of glacier riverwater in winter.

Conclusions

Sub-glacial processes of erosion, transport and deposition, their rates of operation, and the manner in which a glacier moves over its bed are interrelated. Accordingly, landforms fashioned beneath the glacier are influenced by the nature and behaviour of both the ice and the materials over which it moves. Materials at the glacier/bed interface may include debris-rich ice, frozen sediment or rock with interstitial ice, water-saturated unfrozen sediment or rock, and unfrozen, well-drained sediment. As a consequence of the wide range of material properties, the processes which are active beneath glaciers may vary markedly from place to place. Such characteristics of the bed as its irregularity and temperature, the ease with which it deforms, and the manner in which water moves through it affect its interactions with the glacier, whilst deformation of the ice is influenced by its temperature, grainsize and structural anisotropy, and by the air bubbles, sedimentary particles, and other inclusions within it.

Theoretical models of glaciers must be selected with care, particularly in relation to parameters defining basal conditions. A model is a simplification of reality, generally constructed for a particular purpose, and should not be used for any other purpose without careful consideration of its characteristics. Similarly, the results of observations and measurements at particular field sites should not be applied to other situations without caution. Considerable variation in sub-glacial conditions also means that, in the past, glacial processes may have operated at rates different from those observed today.

Finally, glacier motion is not a matter of ice physics alone. Variations in water pressure are also relevant for those glaciologists and geomorphologists who are concerned with glacier sliding, and variations in the dissolved load of outflows reflect modes of water movement in the sub-glacial environment.

References

Andreasen, J.-O. (1983), 'Basal sliding at the margin of the glacier Austre Okstindbre, Nordland, Norway', *Arc. Alp. Res., 15,* 333-8.

Boulton, G.S., Morris, E.M., Armstrong, A.A. and Thomas, A. (1979), 'Direct measurement of stress at the base of a glacier', *J. Glaciol., 22,* 3-24.

Collins, D.N. (1979), 'Quantitative determination of the subglacial hydrology of two Alpine glaciers', *J. Glaciol., 23,* 347-61.

Collins, D.N. (1981), 'Seasonal variation of solute concentration in meltwaters draining from an alpine glacier', *Ann. Glaciol., 2,* 11-16.

Engelhardt, H.F., Harrison, W.D. and Kamb, B. (1978), 'Basal sliding and conditions at the glacier bed as revealed by borehole photography', *J. Glaciol., 20,* 469-508.

Hagen, J.O., Wold, B., Liestøl, O., Østrem, G. and Sollid, J.L. (1983), 'Subglacial processes at Bondhusbreen, Norway: preliminary results', *Ann. Glaciol., 4*, 91-8.

Hantz, D. and Lliboutry, L. (1983), 'Waterways, ice permeability at depth, and water pressures at Glacier d'Argentière, French Alps', *J. Glaciol., 29*, 227-39.

Hooke, R.L. (1981), 'Flow law for polycrystalline ice in glaciers: comparison of theoretical predictions, laboratory data, and field measurements', *Rev. Geophys. Space Phys., 19*, 664-72.

Robin, G. de Q. (1976), 'Is the basal ice of a temperate glacier at the pressure melting point?', *J. Glaciol., 16*, 183-96.

Theakstone, W.H. (1966), 'Deformed ice at the bottom of Østerdalsisen, Norway', *J. Glaciol., 6*, 19-21.

Theakstone, W.H. (1979), 'Observations within cavities at the bed of the glacier Østerdalsisen', *J. Glaciol., 23*, 273-81.

Van de Griend, A.A. and Arwert, J. (1983), 'The mechanism of runoff generation from an alpine glacier during a storm traced by oxygen $^{18}O/^{16}O$'. *J. Hydrol., 62*, 263-78.

2 GLACIAL GEOMORPHOLOGY: TERMINOLOGICAL LEGACY AND DYNAMIC FUTURE

W. Brian Whalley

Introduction

Most visitors to the mountainous parts of the world know that glaciers had some part to play in forming the scenery which they see. For geography students at school, glacial geomorphology is often 'the' geomorphology, or at least that which is most vivid. Unfortunately, as in other branches of geomorphology, the learning of names of features is not very instructive as to *how* things have come to be or, in the usual parlance, about geomorphological processes. It is only in the last few years that information on process studies has been introduced into school textbooks and has become a major part of college texts. Yet the degree to which consideration of process *explains* what has taken place in the landscape is not always clear.

Like a surprising amount of science, the geomorphology taught in school has been known for a long time. However, because it is often so difficult to make good — or even any — observations in glacial geomorphology, incorrect or vague ideas may take a long time to be superseded. This is not to say that all old ideas or observations are wrong: there may not be the opportunity to correct or refine some of these notions. Some of the early pioneers, such as Bonney (1873), can still be read with profit. Glaciology and glacial geomorphology have traditionally gone hand in hand, as can be seen in the classic European texts of Finsterwalder and Machatschek, and in the more recent volumes of Embleton and King (1975), Sugden and John (1976) and Eyles (1983). However, it is possible to detect a divergence of an essentially physics-based glaciology and a more geologically oriented glacial geomorphology. This is evident if one browses through back issues of the *Journal of Glaciology.* The separation can be seen to have begun during the early 1950s. In part, this was due to the need for information about ice behaviour stimulated by the second world war, and more particularly by W.V. Lewis's conviction that the answer to many glacial geomorphological problems could only be found with a knowledge of ice physics. The involvement of J.F. Nye and J.W. Glen, as physicists, into the geographer's world by Lewis in the 1950s provided such a link. However, this ultimately produced specialised papers which most geomorphologists would find difficult to comprehend and may have actually separated glacial geomorphology from a physics base. Through the use of physics, quantification came early to glacial geomorphology. The statistics-based 'quantitative revolution' had quite a different impact in the 1960s and produced numerical morphometry, i.e. description rather than explanation. Perhaps we are now starting to see a convergence of ideas and methodologies, based on examination of materials, properties and the study of mechanisms, and that herein lies an opportunity for progress through a return to the application of the principles of physics. Part of this chapter will be devoted to exploring this idea.

Problems of glacial geomorphology

Peculiar difficulties are faced by the geomorphologist seeking to explain glacial landscape development. This applies in places such as the English Lake District or in the Yosemite valley. Argument is often by analogy; there are no glaciers there at present and so inductive reasoning is paramount. Just as Agassiz persuaded William Buckland in the 1820s that there was truth in the ideas of the glacial theory, so must the modern geomorphologist convince students that what can be seen in front of a glacier in Switzerland or the Cascades could also have happened in Yosemite or a Lakeland valley. To a certain extent this is perfectly justified. However, with this inductive method the analogue may be incorrect, and in addition the testing of the idea (theory, analogue or model — it does not matter) is likely to be difficult. For example, there is clearly a need to observe what is going on at the bottom of present-day glaciers. It has been done, for example, by tunnels (McCall, 1960), with borehole cameras (Koerner *et al.*, 1981) and by natural tunnels (Bennett, 1968), but even these observations may not be representative as only a few sites with specific thermal regimes have been investigated. One of the important ways in which glaciological researchers have impinged upon glacial geomorphology has been the realisation that glacier thermal regimes, particularly at their bases, affect erosion, transport and deposition (Boulton, 1972). Thus Boulton (1972, 1975), amongst others, has discussed the relationship of present-day glaciers as appropriate analogues for Pleistocene icesheets. The traditional idea of Greenland or Antarctica being suitable models has to be modified substantially as more about thermal regimes and their changes through time becomes known.

Much of geomorphology has a temporal viewpoint. Indeed, glacial chronology is, in many ways, often quite separate and distinct from glacial geomorphology, especially as far as process studies are concerned. This separation will be discussed subsequently. The need to answer questions such as 'How fast did it happen?' is obvious, but there is also a requirement for placing events in a relative temporal framework, or at the least in an appropriate sequence. The magnitude frequency question, which is commonly recognised in fluvial geomorphology, is now being applied more specifically to aspects of glacial phenomena. Such investigations clearly rely substantially on the dating of events even though this is often very difficult. The various dating techniques now available may given some success in providing a temporal framework.

In summary, therefore, we can see three basic problems in the elucidation of glacial geomorphological processes:

(a) the need to apply findings from present-day glacier landscapes to those of the past;
(b) the necessity to construct general ideas from often isolated and specific observations by using physical principles; and
(c) the placing of events in the context of time.

Investigations into glacial geomorphology present some problems which may be rather different from other areas of geomorphology. In many parts of the world the glaciers are no longer present, but researchers into presently glacierised areas may have to presume that such glaciers are representative of their appropriate Pleistocene counterparts. Furthermore, the action of ice and sub-glacial meltwater may be largely hidden from view. As well as these difficulties, there are also some problems in interpreting landscapes. For instance, should the terms we use describe landforms and rely on their appearance, or should they imply specific mechanisms, i.e. genesis?

Process and mechanism

The word 'process' has become common in the geomorphological literature over the last ten years or so. It is worthwhile asking what exactly we mean by it. It is useful to think of process as the operation of a mechanism through time. By 'mechanism' we really mean the way in which something works, i.e. the interaction of the various parts of a system. Geomorphology traditionally sees the formation of the landscape in terms of gross changes through time, although 'process studies' have concentrated more and more upon small-scale mechanisms. The events generally studied are small-scale: the removal of a rock fragment by a boulder trapped in basal ice; the opening of a crack by frost action, etc., but the results are usually cumulative and give rise to a description of the process. We can illustrate this with reference to the breakdown of rocks by frost action.

Frost-shattering is a fundamental mechanism for several glacial processes. Examples can be found in ideas on nivation (Thorn, 1979) and the meltwater/bergschrund hypothesis of corrie erosion (Lewis, 1938) and sub-glacial erosion (Bennett, 1968), the production of glacier-transported rock debris (Eyles, 1983) and the pre-weathering of terrain before glaciers incorporate it into the basal debris load (Boyé, 1968).

In *explaining* a geomorphological feature we are faced with two questions: first, what is happening in terms of a mechanism — how is the rock splitting? What are the controls on the action (the internal factors such as rock strength and the external factors such as the rate and intensity of freezing)? Secondly, there are questions relating to the rate of breakdown — for example, how many freeze–thaw cycles must a rock endure before it can be said to have disintegrated? This second type of question, when added to the first, gives a study of process and, overall, provides a dynamic basis for explanation. Unfortunately, examination of the rates of activity alone do not necessarily give an answer to questions relating to process. For instance, Tricart's (1956) study of freeze–thaw cycles used laboratory simulations of 'Siberian' and 'Icelandic' thermal regimes to model frost action in selected rocks (McGreevy, 1981).

By speeding up the thermal cycling he hoped to compress natural activity into a reasonable time-span. Indeed, many other investigators have followed the same line of reasoning. In themselves, however, they are not *process* studies which show a possible rate of breakdown in the field. It might be possible to compare cycles for a specified rock type, or compare rock types undergoing a similar climatic simulation (Lautridou and Ozouf, 1982) but it is not, of itself, a process study. To be particularly critical, one could also say that this type of experiment is not a study of mechanism either, but one of investigating controls on activity. In fact, many field experiments are of this type too. For an understanding of what actually happens, and how the various internal and external controls *interact*, then the experimentation must be very much more detailed. To make a more complete experiment necessitates the detailed examination of what happens mechanically to the rock investigated. In other words, the experiment starts to become a part of physics or engineering.

The other assumption made by workers on frost-shattering of rocks is that breakdown is controlled by the rate of freezing, or its intensity. The significance of water for such experiments may be crucial, but its position in cracks or pores in relation to the freezing front is largely unknown. White (1976) suggested that what is often called frost-shattering may be due to the presence of water alone. This idea stems from work on limestone by Dunn and Hudec (1966); how well it can be applied to other rocks remains to be seen. Although White may have overstated the case for 'hydration shattering' his scepticism generally is well placed. Indeed, the idea that a succession of air temperature changes can be used to infer rock breakdown is itself in question (McGreevy and Whalley, 1982). The way in which the rock reacts to temperature changes — the external controls — in the air is the significant factor, not the air temperature fluctuation itself (Douglas *et al.*, 1983). The internal controls — porosity, crack density, tensile strength, etc. — all need to be taken into consideration in any understanding of the mechanism of rock breakdown.

Some would argue (Chorley, 1978) that it is not the role of the geomorphologist to study detailed mechanisms but that this should be left for others — physicists, chemists, engineers, etc. This view is, to some extent, justified. Nevertheless, if the overall geomorphic controls are to be evaluated, then it does become necessary to search for the detailed explanation as well as identify the overall results in the landscape. There is scope for both; indeed, there must be both if attempts to make better explanations of geomorphological features are to succeed.

Terminology and the study of process

A problem pervading much of glacial geomorphology is that of terminology. The trouble is insidious in its effects rather than a concern which is frequently in the forefront of research. At one level, the use of certain terms belies a knowledge of what goes on. Few fluvial geomorphologists would have difficulty in saying what a meander is; that is, what it looks like and how one might describe it. Similarly, a glacial gemorphologist would be able to describe a drumlin. Morphologically there is little problem in either case. However, whereas there is plenty of information about meander geometries, relationships to rivers, bank material, erodibility mechanisms and rates, etc. as well as a good mechanical and theoretical base for explaining why meanders are as they are, the same can hardly be said of drumlins. There are theories (Smalley and Unwin, 1968; Menzies, 1979a) as well as varied descriptions (see e.g. Menzies, 1979b), but the observational problems mentioned previously present a shortage of knowledge about this and many glacial landforms. The very fact that a name is given to a feature may tend to suggest that we know more about it than we really do. As well as drumlin we could place frost-shattering (McGreevy, 1981), the 'flutes' of Figure 2.1 (Paul and Evans, 1974; Boulton, 1976), and 'eskers' (Embleton and King, 1975) in a similar category. Of course, things are known about these phenomena but perhaps because the names are of such long standing, this frequently belies understanding. This is true from the viewpoint of both mechanisms and of processes. No amount of description of the land*form* will help elucidate the formational mechanism. With drumlins for instance, the classic and often-cited paper by Chorley (1959) is concerned with such quantitative description and not the mechanics of formation (Boulton, 1976).

Problems can arise if a term becomes synonymous with a mechanism or process. An idea maybe difficult to displace if it needs to be modified or replaced and this may mean that one model dominates the 'explanation' to the exclusion of others. At one time the presence of erratics was thought to signify the effects

Figure 2.1: Fluted ground moraine, or 'flutes', Northern Norway. The surface is covered with these ribs of moraine material, rarely more than 50 cm high, left by a receding glacier.

of the Biblical Flood. The overthrow of the 'Diluvian theory' was difficult because the idea of glaciers transporting them was so revolutionary (Chorley *et al.*, 1964).

If a unique mechanism is associated with a single topographic form we can call it a genetic explanation. Haines-Young and Petch (1983) have argued that it is necessary to have such a classification of landscape features, but this does have some disadvantages. We are here in the realms of philosophy of science and psychology as much as geomorphology. Nevertheless, these considerations are important because they affect both the learning of the subject and the way in which research is carried out.

The bergschrund or meltwater hypothesis of corrie erosion (Lewis, 1938), which requires the freezing of water on the exposed rock surface near the base of a bergschrund in response to air temperature fluctuations, certainly needs modification in the light of investigations by Battle (1960) and McGreevy and Whalley (1982), amongst others. White (1976) has even suggested that 'freeze–thaw' is not so much volumetric expansion of water but 'hydration shattering'. Although this may be an extreme view and can be disputed, there is always the danger that hopping on a bandwagon can create one terminological problem from another by substituting 'hydration shattering' for 'volumetric expansion of water' in any explanation. One paper may exert a great influence on those who follow, but the direction may not, in retrospect, always be the best one. Unfortunately, this seems to be an unavoidable consequence of the way in which knowledge usually builds upon what has gone before. It means that the direct linking of a landform with a cause or explanation should be treated with some caution and that a questioning approach is desirable.

The problem of genetic implications just mentioned is an important one. Some workers would argue that a term needs a genetic connotation — this is the view of Haines-Young and Petch (1983) — i.e. the term itself denotes the mechanism or process involved. Thus, it is clear from the use of 'meander' that there is a river involved and that erosion of the bank material at certain places accentuates the form (see e.g. K.S. Richards (1982) for further discussion). A word like 'avalanching' means quite different mechanisms according to usage. For the glaciologist or snow hydrologist ideas concerning the temperatures and moisture content of snow layers come to mind. The same term is used by sedimentologists in discussing the movement of dunes or deltas by the progressive rolling or sliding of sandy sediment down a lee slope. In both cases there is a non-genetic component, that of the falling, rolling or sliding material, snow or sand, but there is an implicit link with the mechanism by which the movement has been started. This is clearly different in each case. The context may supply the clue here but this is not always clearcut. We can find similar links between landforms and mechanisms for terms in glacial geomorphology.

One example of this is the discussion of the formation and flow mechanisms of rock glaciers. For some authors (e.g. Barsch, 1977) it is implicit that these features are formed as a mixture of snow, ice and refrozen meltwater and that the ice formed lies in the interstices of rock debris which constitutes the rock glacier. Thus any mention of 'rock glacier' implicitly means this structure and flow mechanism. Others would suggest that there are problems with this mechanical model (Whalley, 1974; Washburn, 1979) and that other ideas need to be investigated. The decision as to which model of flow mechanics is most appropriate requires the application of physical and engineering principles: this is not helped if a term is exclusively used to denote one, perhaps equivocal, model. Thus it is preferable that the term 'rock glacier' be used just to describe the feature rather than to infer a particular mechanism.

The point of giving something a name is basically to allow communication in a succinct way. Primarily, it must be as unambiguous as possible so that it can be understood by others. Most often this is done by reference to some form of sense data — what it looks like, its colour, etc. In some cases this is easy; it should not be difficult for someone to recognise and differentiate betwen a soccer ball and a golf ball on the basis of a verbal description even if they had not known what they looked like before. With a rock glacier, this is not quite as easy. However, even if one had not seen one before, looking at Figure 2.2 would probably give a basic idea sufficient to enable recognition of other examples unaided. It is not necessary to know how they are formed for this identification procedure, and it is then possible to supplement pictorial information with specific details such as 'They are composed, at least on the surface, by rock debris which has fallen from mountain sides; they are found in many high mountain parts of the world; they have a basic morphology similar to that of glaciers, show flow features but generally have surface velocities and order of magnitude less than true glaciers.' These additional data are again descriptive and do not presume much in the way of genetic implication and nothing about their flow mechanics. Thus, it is possible to say that if an object looks like a rock glacier (or any other feature) then it can be called that. However, having made the identification, it is then possible to investigate the feature under study and suggest models of origin and formation, environmental implications, etc. But if the term used presupposes these considerations, then it is difficult to break away from that strait-jacket. Rock glaciers and frost-shattering are just two examples of many that are used in geomorphology. In some cases, the feature may be understood well enough to allow a substantial genetic implication, but as there is so little known about many glacial landforms there is danger in using genetic terminology too freely.

The idea that a glaciated valley has a U-shaped cross-profile is an old one which goes back at least to an article by W.M. Davis in 1900: 'the glacial channel is U-shaped, broad and deep, while the valley flares open, V-like, above the ice surface' (Davis, 1954). It seems almost conventional to say that valleys which once contained glaciers (glaciated valleys) have U-shapes and that valleys

Figure 2.2: A rock glacier in the Wrangell Mountains, Alaska. The form is distinctive and can be recognised in many mountains of the world. Despite some 70 years of observations and investigations on the origin and formation of features such as this, we are still some way from achieving a consensus view.

which did not contain glaciers have V-shapes. This is a simplification which is both confusing and untrue. Svensson (1959) asked, 'Is the cross-section of a glacial valley a parabola?' It may well be described by a parabola but this could be contested as strongly as the U-shape idea. There are certainly valleys which have V-shapes and interlocking spurs and which have contained glaciers within the last few thousand years — the classic examples are Lauterbrunnen, Switzerland, and Yosemite, USA, which are exceptional cases of U-shapes with very steep sides, being much more pronounced than the cross-profiles seen in many other mountain areas. These examples are related to peculiarities of rock type and tectonics. A recent note by Small (1982), particularly useful in the context of glacial action, also points out that many definitely glaciated valleys do not have U-shapes. This is clearly another case of form being too closely identified with process.

As well as differences in form, we can also consider similarities. The concept of 'equifinality' (or 'convergent landforms') is, roughly speaking, the idea that similar landforms can be produced in different ways (Pitty, 1982). Haines-Young and Petch (1983) suggest that the idea can be used in three ways:

(1) because similar forms do result from diverse origin in different ways;
(2) because the understanding of the process is deficient and mechanisms which are apparently dissimilar are in fact the same; or
(3) because the forms only appear to be similar when in reality they are different and so reflect their diverse origins.

Haines-Young and Petch (1983) believe that the last two are incorrect. It is not possible here to discuss all the ramifications of these ideas, but it is clear that when there is doubt about the origin of features then very specific investigations are necessary to determine if equifinality (as in (1)) is possible. However, it is possibly the case that use of the concept hides ignorance, not quite in the sense of (2), but because it is not even possible to decide which of the three meanings is relevant. Non-genetic terms and classifications could well lead into the trap of ascribing incorrect mechanisms to certain landforms. Nevertheless, it should be realised that, because of the uncertainty in knowing much about formative mechanisms of many geomorphological features, it is expedient to concentrate on a study of mechanism rather than form.

Time in glacial geomorphology

One of the fundamental difficulties with many aspects of glacial geomorphology is the length of time many features take to form. This is evident in various ways, the most obvious being that glaciers grow only slowly when there is a response to climate change sufficient to induce glacier growth or extension. The 'response time' — that is, the time taken for a change in mass balance to come to equilibrium in the system — may be of the order of hundreds of years. Even the 'lag time' (the time taken for a change in mass balance to be seen at the snout of the glacier) may be several years. Thus, although there are a few long-term records in glaciology (e.g. Haefeli, 1970) there is nothing which really corresponds to this when we look at the effects of glaciers on the landscape. It may, of course, be possible to make inferences regarding long periods, indeed, this is one of the applications of establishing glacial chronologies for an area. However, there are inbuilt difficulties in this since even if the start and finish of a particular episode of activity are known it may not be easy (or even possible) to say what the rate of activity was, and whether it remained constant throughout the period investigated. This is partly tied into current investigations as to the magnitude and frequency of geomorphic events.

If we can date the start of glacial activity in a valley as well as its end, and want to evaluate the role of glacier erosion, then we also need to know the *initial* shape of the valley: to assume that the valley was V-shaped prior to glacierisation is clearly insufficient; we also need to know how deep the valley was. One of the classic examples of this kind of investigation is that of Matthes in the Yosemite valley, California (Matthes, 1930; Sugden and John, 1976). On a rather smaller scale is the work by Jahns (1943) on the removal of blocks from the top of an outcrop showing sheeting structures. On the whole, however, this type of investigation is rather rare. Only occasionally is it possible to carry out experiments which look at the erosion of material by glaciers directly in the lifetime of an investigator. For example, even if we can estimate the glacial erosion of a valley system how do we know when exactly the activity occurred? Was it during an intense period when thermal conditions were particularly favourable, or was it a slow continuous process? It may be possible to make intelligent guesses at answering such questions but they may only be just that because of the difficulties in interpreting the results of sampling over a short period placed within a much larger time period (Thornes and Brunsden, 1977, p. 61). This, of course, will always remain a problem but the investigator must at least be aware of it and try to design experiments which probe it.

Glacial chronology

The establishment of glacial chronologies can become separate from the aspects of glacial geomorphology which investigate processes and mechanisms. Such studies may be an end in themselves: an area to be examined, mapped and dated and which may, under favourable circumstances, be related to neighbouring areas. Indeed, with the increasing need to examine climatic trends, linking such studies globally has become important with such projects as CLIMAP. However, this is not to say that global correlations are easy. It is still difficult to place the Quaternary chronology of many areas (e.g. the British Isles) on a fully correlated basis (Mitchell *et al.*, 1973). Until recently, it was thought that there were no sites dating from the Last Interglacial in Ireland (Herries Davies and Stephens, 1978) because deposits which stratigraphically might be expected to be of this date were found to contain plant material similar to the Penultimate (Hoxnian) Interglacial in East Anglia. The inference was thus made that the Irish material was of similar age to that in East Anglia — this, despite the fact that the English Midlands cannot be correlated with East Anglia, only 100 km away! It is more likely that the true Last Interglacial deposits in Ireland are not 'lost' nor lie deeply buried, but are in fact those presumed to be of Penultimate Interglacial age. In New Zealand, moraines which were once thought to have been deposited in the last 800 years have now been dated from before 6000 years BP to the present (Gellatly, 1984). This revision was the result of using a 'multi-parameter' dating approach as advocated by Birkeland (1982) rather than a single criterion. An interesting comment on glacial chronologies is presented by Birkeland *et al.* (1979) in which they look at the problems concerned with relative dating techniques and the way in which such information can be used.

Glacier behaviour is not necessarily synchronous in any one area. Nevertheless, in Norway the prominent moraine, which in the South is known as the '1750 moraine' (i.e. 1750 AD), has been 'identified' in areas over a 1000 km away. Whilst there *may* be a correlation, accurate dating is still required to confirm this event. The various pitfalls briefly reported here should not occur if more attention is taken of glacial processes. Even a multiplicity of radiocarbon dating to assist correlations is not necessarily the answer. A large number of dates from moraines of glaciers in the Southern Alps of New Zealand have served to compound the complexities of the glacial history rather than clarify it (Gellatly *et al.*, in press).

There are many ways in which the dating of glacial events is extremely important for the glacial geomorphologist. If it is necessary to place events in sequence, particularly when looking at rates of activity or process, it is essential. Such endeavours are well shown by the work of Andrews in using raised shoreline reconstructions to study post-glacial land uplift (Andrews, 1970).

A very clear link is forged between glacial geomorphological investigations into mechanisms of deglaciation and glaciology by the work of Sissons and Sutherland (1976), in which they attempt to reconstruct the Pleistocene ice mass for Scotland by consideration of ice limits. These limits, taking the form of moraines, etc., have been found over several years' work by field observations. Unfortunately, such studies are very time-consuming, especially in the field, yet they do promise an important link between the reconstruction of past climates and glacial processes. Having mapped the glacier limits for an area and placed them in a temporal sequence, the authors then use ideas on glacier dynamics to reconstruct the possible ice mass for the South-east Grampians of Scotland and make inferences about the climate for a period in the Late Glacial (Loch Lomond re-advance=Younger Dryas). The model they use involves computing a regional snow line from knowledge of accumulation data based on Norwegian glaciers. The Norwegian model used by these authors uses data from South Norwegian valley glaciers as a basis. In the Lyngen Peninsula in North Norway, there are small icecaps on mountain tops (Figure 2.3) as well as valley glaciers. Although both types of glacier are retreating, the very existence of the substantial amounts of ice on the summit plateaux, similar in topography to the mountain plateaux of Central Scotland, affects the behaviour of the valley glaciers. It is the valley glaciers which produce moraines, not the plateau tops. However, the valley glaciers in Lyngen do not get most of their ice from snowfall in the usual way (e.g. Sugden and John, 1976) but from the avalanching of ice from the summit icecaps into the valleys. The model so far used for Scotland does not take such a separation of plateau icecaps and valley glaciers into account. This could affect predictions about the climate as inferred from moraine positions and chronology more than has been realised.

The interesting analysis of Sissons and Sutherland (1976) provides much food for thought, as does the related work of Gordon (1979), who used a reconstructed icesheet for Northern Scotland to examine the effects of glacier erosion, and that of Sugden (1977) for the Laurentide icesheet. A good review of the general progress in icesheet reconstruction is given by Andrews (1982).

Figure 2.3: The edge of an icecap on a mountain top (1700 m) in Northern Norway. The ice is slowly receding from the edge of the plateau, but the majority of ice in the area is on the mountain tops and not in the valleys. The icecaps tend to feed the valley and corrie glaciers by avalanching, and the valley glaciers probably receive very little ice mass from direct snowfall.

Materials and glacial geomorphology

A knowledge of materials and the way they behave is an important aspect of geomorphology (Whalley, 1976). Thus, quantitative geomorphology should be concerned with the physical and chemical aspects of any problem rather than statistical morphometry if the mechanisms, and not just the controls of operation, are to be explored. Then, having some knowledge of the mechanisms, it is possible to look into time-related aspects with more clarity. This may only be an ideal, but it should help to place geomorphological studies on a better footing and allow them to benefit from theories presented elsewhere. Such ideas are especially pertinent to glacial geomorphology, as would be expected given the close relationship between glaciology and glacial geomorphology.

One of the examples used previously, that of frost-shattering of rocks, shows how traditional ideas can be extended fruitfully by looking at material properties. There is not a great deal of theory

Figure 2.4: An experimental rock weathering site at 4400 m in the Karakorum Mountains, Pakistan. A block of rock with properties tested in the laboratory is surrounded by expanded polystyrene (to avoid edge effects) and examined at the same time as the local rock.

concerning this phenomenon although recent work by Hallet (1983) has shown how a study of ice lensing and heave in frozen soils can be used to examine rock fracture. Such work depends upon physical principles using crack propagation theory with equilibrium conditions set by an equation which determines the freezing-point depression with applied pressure. Once there is a physical basis for a theory in this way then it is possible to devise experiments which test the model in the field or laboratory. No doubt there will be the chance to fit in the environmental controls derived from the experiments in the 'classic' Tricart manner, but the real progress is in defining a model which rests upon a sound physical basis. This is but one example where the introduction of relatively simple physics to glacial geomorphology will help overcome the inbuilt limitations of many so-called process experiments. At the same time, there is a need to collect good field data to enable laboratory testing to be done with a reasonable representation of reality. The problems of using air temperatures as surrogates for rock temperatures has already been noted. One technique

(Figure 2.4), is to examine the temperatures on rocks in the field but at the same time subject specimen blocks to the same conditions (Whalley *et al.*, 1984). The test blocks are investigated fully in the laboratory and all the requisite properties (thermal conductivity, specific heat capacity, surface reflectance, etc.) determined so that a link is provided between field and laboratory experimentation.

One further illustration is the way in which ideas on the formation of glacial flutes (fluted ground moraine, Figure 2.1) have changed from simple description (e.g. Dyson, 1952) through the use of soil mechanical principles (Paul and Evans, 1974) to a much more complex analysis by Morris and Morland (1976).

The necessity to examine such problems from the point of view of the mechanics of the materials provides new insights into a relatively common glacial feature. The discussion about the origin of rock glaciers, touched on briefly above, is also amenable to specific investigation by glaciological theory (Whalley, 1974).

Conclusions

Glacial geomorphology is starting to apply physical principles to a variety of long-standing problems. This will not replace the conventional, field-oriented approach but modify it so that field investigation becomes more experimental, perhaps increasingly cross-linked to experimental laboratory work. This can only be to the good, but it does presuppose a willingness by geomorphologists to learn the necessary theory and apply it. It will certainly not be easy for traditionally trained geomorphologists and geologists to cope with increasingly complex analyses of landforms. Nevertheless, if we are to learn more about how things have come to be as they are then investigation of mechanisms will become important.

References

Andrews, J.T. (1970), 'A geomorphological study of post-glacial uplift, with particular reference to Arctic Canada', *Inst. Br. Geographers,* Spec. Publ., *2.*

Andrews, J.T. (1982), 'On the reconstruction of Pleistocene ice sheets: a review', *Quat. Sci. Rev.*, *1,* 1-30.

Barsch, D. (1977) 'The nature and importance of mass wasting by rock glaciers in alpine permafrost environments', *Earth Surface Processes,* 2, 231-45.

Battle, W.R.B. (1960), 'Temperature observations in bergschrunds and their relationship to frost shattering', in Lewis, W.V. (ed.), *Investigations on Norwegian Cirque Glaciers, Royal Geographical Society research series, 4,* 83-95.
Bennett, R.G. (1968), 'Frost shatter and glacial erosion under the margins of Østerdalsisen, Svartisen', *Norsk Geografisk Tidsskrift, 22,* 209-13.
Birkeland, P.W. (1982), 'Subdivision of Holocene glacial deposits, Ben Ohau Range, New Zealand using relative-dating methods', *Geol. Soc. Am. Bull., 93,* 433-49.
Birkeland, P.W., Colman, S.M., Burke, R.M., Shroba, R.R. and Meierding, T.C. (1979), 'Nomenclature of alpine glacial deposits, or, what's in a name?', *Geology, 7,* 532-6.
Bonney, T.G. (1873), 'Lakes of the North-Eastern Alps and their bearing on the glacier erosion theory', *Q. J. Geol. Soc. Lond., 29,* 382-95.
Boulton, G.S. (1972), 'Modern arctic glaciers as depositional models for former ice sheets', *J. Geol. Soc. Lond., 128,* 361-93.
Boulton, G.S. (1975), 'Processes and patterns of subglacial sedimentation, a theoretic approach', in Wright, A.E. and Moseley, F. (eds.), *Ice Ages Ancient and Modern,* (Seel House, Liverpool), 7-42.
Boulton, G.S. (1976), 'The origin of glacially-fluted surfaces — observations and theory', *J. Glaciol., 17,* 287-309.
Boyé, M. (1968), 'Défense et illustration de l'hypothèse du défonçage periglaciaire', *Biuletyn Peryglacjalny, 17,* 5-56.
Chorley, R.J. (1959), 'The shape of drumlins', *J. Glaciol., 3,* 339-44.
Chorley, R.J. (1978), 'Bases for theory in geomorphology', in Embleton, C., Brunsden, D. and Jones, D.K.C. (eds.), *Geomorphology, Present Problems and Future Prospects* (Oxford University Press, Oxford), 1-13.
Chorley, R.J., Dunn, A.J. and Beckinsale, R.P. (1964), *The History of the Study of Landforms,* vol. 1 (Methuen, London).
Davis, W.M. (1954), 'Glacial erosion in France, Switzerland and Norway', *Geographical Essays* (Dover, New York), ch. 24.
Douglas, G.R., McGreevy, J.P. and Whalley, W.B. (1983) 'Rock weathering by frost shattering processes', *Proc. 4th Int. Conf. Permafrost* (National Academy Press, Washington), 244-8.
Dunn, J.R. and Hudec, P.P. (1966), 'Water, clay and rock soundness', *Ohio J. Sci., 66,* 153-68.
Dyson, J.L. (1952), 'Ice-ridged moraines and their relation to glaciers', *Am. J. Sci., 350,* 204-12.
Embleton, C. and King, C.A.M. (1975), *Glacial Geomorphology* (Arnold, London).
Eyles, N. (1983), *Glacial Geology* (Pergamon, Oxford).
Gellatly, A.F. (1984), 'The use of rock weathering-rind thickness to re-date moraines in Mt. Cook National Park, New Zealand', *Arctic and Alpine Res., 16,* 225-32
Gellatly, A.F., Rothlisberger, F.R. and Geyh, M. (in press), 'Radiocarbon dating and moraine chronologies for New Zealand', *Z. Gletscherkunde and Glazialgeol.*
Gordon, J.E. (1979), 'Reconstructed Pleistocene ice sheet, temperatures and glacial erosion in northern Scotland', *J. Glaciol., 22,* 331-44.
Haefeli, R. (1970), 'Changes in the behaviour of the Unteraargletscher in the last 125 years', *J. Glaciol., 9,* 195-212.
Haines-Young, R.H. and Petch, J.R. (1983), 'Multiple working hypotheses: equifinality and the study of landforms', *Trans. Inst. Brit. Geographers, 8,* 458-66.

Hallet, B. (1983), 'The breakdown of rock due to freezing: a theoretical model', *Proc. 4th Int. Conf. Permafrost* (National Academy Press, Washington), 433-8.

Herries Davies, G.L. and Stephens, N. (1978), *Ireland* (Methuen, London).

Jahns, R.H. (1943), 'Sheet structure in granite: its origin and use as a measure of glacial erosion in New England', *J. Geol., 51,* 71-98.

Koerner, R.M., Fisher, D.A. and Parnandi, M. (1981), 'Bore-hole video and photographic cameras', *Ann. Glaciol., 2,* 34-8.

Lautridou, J.P. and Ozouf, J.C. (1982), 'Experimental frost shattering: 15 years of research at the Centre de Géomorphologie du CNRS', *Progress in Physical Geography, 6,* 215-32.

Lewis, W.V. (1938), 'A meltwater hypothesis of cirque formation', *Geol. Mag., 75,* 248-65.

Matthes, F.E. (1930), 'Geologic history of the Yosemite valley', *US Geol. Survey Prof. Pap., 160.*

McCall, J.G. (1960), 'The flow characteristics of a cirque glacier and their effect on glacial structure and cirque formation', in Lewis, W.V. (ed.), *Investigations on Norwegian Cirque Glaciers, Royal Geographical Society research series, 4,* 39-62.

McGreevy, J.P. (1981), 'Some perspectives on frost shattering', *Progress in Physical Geography, 5,* 56-75.

McGreevy, J.P. and Whalley, W.B. (1982), 'The geomorphic significance of rock temperature variations in cold environments: a discussion', *Arctic and Alpine Res., 14,* 157-62.

Menzies, J. (1979a), 'The mechanics of drumlin formation with particular reference to the change in pore-water content of the till', *J. Glaciol., 22,* 373-84.

Menzies, J. (1979b), 'A review of the literature on the formation and location of drumlins', *Earth Sci. Rev., 14,* 315-59.

Mitchell, G.F., Penny, L.F., Shotton, F.W. and West, R.G. (1973) 'A correlation of Quaternary deposits in the British Isles', *Geol. Soc. Lond., Special Report, 4.*

Morris, E.M. and Morland, L.W. (1976), 'A theoretical analysis of the formation of glacial flutes', *J. Glaciol., 17,* 311-23.

Paul, M.A. and Evans, H. (1974), 'Observations on the internal structure and origin of some flutes in glacio-fluvial sediments, Blomstrandbreen, North-West Spitsbergen', *J. Glaciol., 13,* 393-400.

Pitty, A.F. (1982), *The Nature of Geomorphology* (Methuen, London).

Richards, K.S. (1982), *Rivers, Form and Process in Alluvial Channels* (Methuen, London).

Sissons, J.B. and Sutherland, D.G. (1976), 'Climatic inferences from former glaciers in the South-East Grampian Highlands, Scotland', *J. Glaciol., 17,* 325-46.

Small, R.J. (1982), 'Glaciers: do they really erode?', *Geography, 67,* 9-14.

Smalley, I.J. and Unwin, D.J. (1968), 'The formation and shape of drumlins and their distribution and orientation in drumlin fields', *J. Glaciol., 7,* 377-90.

Sugden, D.E. (1977), 'Reconstruction of the morphology, dynamics and thermal characteristics of the Laurentide ice sheet at its maximum', *Arctic and Alpine Res., 9,* 21-47.

Sugden, D.E. and John, B.S. (1976), *Glaciers and Landscape* (Arnold, London).

Svensson, H. (1959), 'Is the cross-section of a glacial valley a parabola?', *J. Glaciol., 3,* 362-3.

Thorn, C.E. (1979), 'Bedrock freeze–thaw weathering regime in an alpine environment', *Earth Surface Processes, 4,* 211-28.

Thornes, J.B. and Brunsden, D. (1977), *Geomorphology and Time* (Methuen, London).

Tricart, J. (1956), 'Étude experimentale du problème de la gelivation', *Biuletyn Peryglacjalny, 4*, 285-318.
Washburn, A.L. (1979), *Geocryology* (Arnold, London; 1980 edn, Wiley, New York).
Whalley, W.B. (1974), 'Rock glaciers and their formation as part of a glacier debris-transport system', *Geographical Papers* (Reading), *24.*
Whalley, W.B. (1976), *Properties of Materials and Geomorphological Explanation* (Oxford University Press, Oxford).
Whalley, W.B., McGreevy, J.P. and Ferguson, R.I. (1984), 'Rock temperature observations and chemical weathering in the Hunza region, Karakorum: Preliminary data', in Miller, K.J. (ed.), *International Karakorum Project,* Vol. 2, 616-33.
White, S.E. (1976) 'Is frost action really only hydration shattering?', *Arctic and Alpine Res., 8*, 1-6.

3 ACTIVE AND FOSSIL PERIGLACIAL LANDFORMS

Anders Rapp

Introduction

Sweden covers a long North–South strip of land, from about latitude 55°N in the southernmost province of Scania to 69°N in the northern tip of Lappland. South Sweden has a temperate climate with annual mean temperatures about +7°C (Malmö +7.7°C), and mild winters. Northern Lappland has annual mean temperatures of about −1°C at low levels (Kiruna −1.9°C). The northern mountains of Lappland, with glaciers and vast areas of mountain tundra, have attracted generations of scientists to study actual conditions of glacial and periglacial environments. One obvious reason for these studies has been to make comparisons with landforms and other traces from periods in the past, when the large Quaternary glaciations or the surrounding tundra zones were also active in the far South of the country, and made its characteristic imprint on the landscape there. The study of how present or past climates result in typical landforms which can be used as indicators of climatic conditions is called climatic geomorphology. Such comparative studies are still only beginning, and much further work is needed before thorough interpretations of the full impact of periglacial conditions on the landscape of southernmost Sweden can be made.

Until recently it was the general belief that South Sweden was entirely dominated by landforms from the latest glaciation — the Weichselian — or its deglaciation, with strong impact of glacial meltwater and associated coastal wave action. In recent years, a modified view has emerged, with more emphasis on the periglacial modifications of the landscape, created during past tundra periods after the deglaciation of the Weichselian icesheet, and formed during very long tundra periods before the Weichselian icefront arrived from the North (Figure 3.1).

This chapter briefly presents some periglacial indicator landforms which are being studied in the North and South of Sweden. They are, first, ice-wedge casts and frost crack polygons, and second, nivation hollows and other effects of nivation processes.

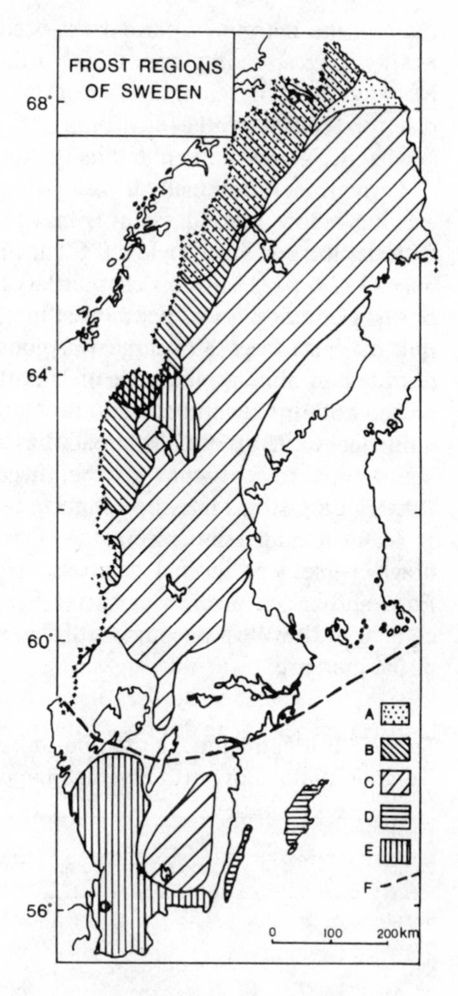

Figure 3.1: Regional division of Sweden according to the types of frost phenomena in the ground. A. Region with discontinuous permafrost on low levels, mainly palsa bogs. B. Caledonide mountains. C. Region of boulder depressions. D. Islands of Öland and Gotland. E. Region with fossil ice-wedges. F. Younger Dryas icefront re-advance 10 200 BP (modified after Lundqvist (1962) and Johnsson (1983)).

Contemporary periglacial environment in Northern Lappland

Lundqvist (1962) summarised the known observations of periglacial features in a map of the 'Frost regions of Sweden'. In the north-western corner of the country is an area mapped as 'region with sporadic permafrost on low levels', coinciding with the Caledonian mountain range and its foreland to the East. This is the area mapped as the outermost extension westwards of the Eurasian belt of 'discontinuous permafrost' (Washburn, 1979, p. 25). The northern Swedish area of discontinuous permafrost is thus on the margin of permafrost occurrence in Eurasia, like the nearby corresponding zones in Northern Finland and Northern Norway. As a marginal area of permafrost it is of particular interest for studies of the environmental factors which control the occurrence of permafrost and its fluctuations in time.

Permafrost (perennially frozen ground) is defined as a condition existing below ground surface, in which the temperature in the material has remained below 0°C for more than two years (Washburn, 1979, p. 21). In this chapter large polygons will be discussed as surface indicators of permafrost in Northern Sweden. The plotting of palsa bogs and large polygons is based mainly on geomorphological maps at a scale of 1: 250 000 (Melander, 1977) and on the author's own observations. The mapped area (Figure 3.2), more accessible in 1983 as a road, has been built close to the existing railway from Kiruna to the Swedish–Norwegian border at Riksgränsen. It will later be connected with Narvik on the coast.

Mean annual temperatures and precipitation for weather stations (Table 3.1) show a decrease in precipitation from West to East, and also a drop in mean temperatures, as a more maritime climate in the West becomes more continental in the eastern part of the map area.

Table 3.1: Mean annual air temperatures and precipitation for weather stations of north Swedish Lappland

Station	Altitude m asl	Temperature	Precipitation	Period
Riksgränsen	522	−1.5°C	844 mm	1901-30
Abisko	388	−1.0°C	267 mm	1901-30
Kiruna	510	−1.9°C	453 mm	1901-30

Most present-day cirque glaciers as well as empty glacial cirques in the mountains of Lappland face towards the East and North-east (Figure 3.3). Similarly, present-day perennial snow patches in the Abisko region mainly face East and North-east (Figure 3.3c). Many other signs of the effects of predominant wind directions in winter agree with the pattern of mainly westerly snow drifting. This is particularly strong above the tree limit, in open valleys and on open plateau surfaces.

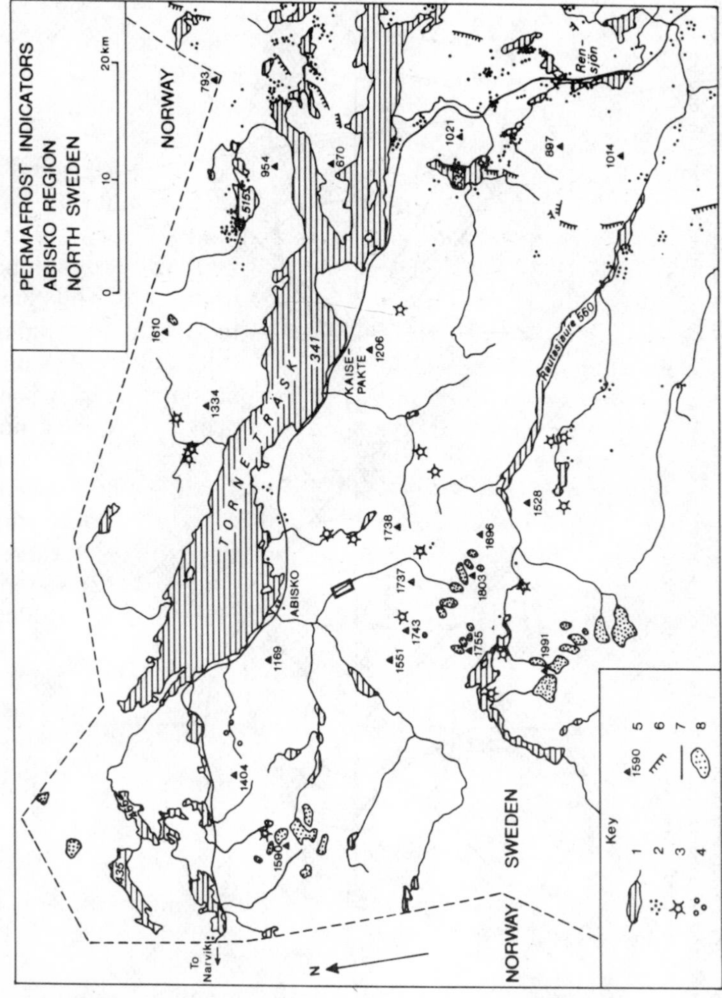

Figure 3.2: Map of permafrost indicators in the Abisko region, North Sweden. Grey tone indicates areas above 600 m asl, which is the approximate limit for birch forest in this region. Key: 1 — lake. 2 — palsa bogs. 3 — large frost-crack polygons. 4 — thermokarst ponds (small 'collapsed pingos'). 5 — mountain summit and altitude in metres. 6 — supposed neotectonic faults. 7 — motor road. 8 — glacier. (Based on Melander (1977), modified by Rapp.)

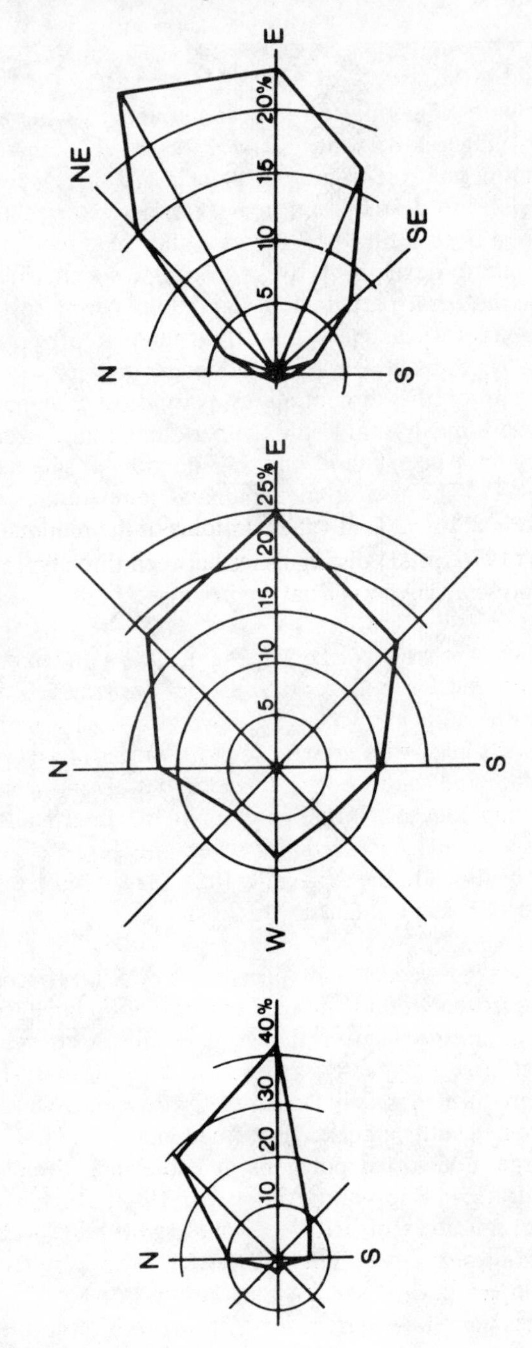

Figure 3.3: The orientation of present-day glaciers (a), glacial cirques (b), and perennial snow patches (c) in the Swedish northern mountains. They all show a marked predominance of E and NE aspects, and thus a predominant direction of snow-drifting by winds from the westerly quadrant during the whole period of formation of cirques. (a) represents 237 cases, (b) 1227 cases and (c) 592 cases. (From Vilborg (1977).)

Ice-wedge casts and frost-crack polygons in Lappland

Palsas are lenses of permafrost in peat and underlying mineral soil. Palsa bogs (Figures 3.1 and 3.2) are regarded as indicators of rather weak permafrost, requiring a mean annual temperature of −1°C or colder to develop and persist. Another surface indicator of permafrost is ice-wedge polygons, but they require a much more severe climate to develop. Actively growing ice-wedges in Alaska 'occur, for the most part, in the continuous permafrost zone of northern Alaska where mean annual air temperatures range from −6° to −12°C.... Inactive wedges, no longer growing, occur in the northern part of the discontinuous permafrost zone of central Alaska, where mean annual air temperatures range from −2° to −8°C; they have been found only in fine-grained silty sediments' (Péwé, 1964). The lowest mean annual temperatures in north Sweden are −2° to −3°C in valley bottoms of the mountains.

French (1976, p. 21) distinguishes between three types of frost-cracks in present-day periglacial environments.

> First, there are frost cracks which are filled with ice. These are called ice wedges.... Second, there are frost cracks, which are filled with sand, known as sand wedges.... Third, there are frost cracks filled with mineral soil known as soil wedges. While ice wedges and sand wedges develop not only in the seasonally thawed zone but also in the perennially frozen ground beneath, and may extend downwards for several metres, soil wedges are usually confined to the seasonally frozen layer and rarely extend downwards for more than 1.0 to 2.0 m.

Ice-wedges, sand-wedges and soil-wedges develop by contraction cracking in frozen ground and they correspond to large frost-crack polygons on the surface of the ground. Tephrochronology has shown that large polygons developed in non-permafrost areas in Iceland were formed mainly during the Little Ice Age in the seventeenth to nineteenth centuries (Friedman *et al.*, 1971).

The large, non-sorted polygons in Padjelanta, Swedish Lappland (Figure 3.4; Rapp and Annersten, 1969; Rapp and Clark, 1971) are indicators of actual permafrost on windswept tundra areas without peat cover. They have small ice-wedge casts under well-developed podzol soil profiles below polygon furrows. At some places they show vertical ice-filled fissures of maximum 1 cm

Figure 3.4: Large frost-crack polygons and inclined, non-sorted steps on windswept ridges of till, Padjelanta National Park, Lappland. Predominant winds from upper left to lower right corner. (From Rapp and Clark (1971).)

width in permafrost down to at least 2.15 m below the ground surface. In the active layer, which is about 1-1.5 m thick, the wedge casts have a reddish fill of a podzol B horizon, indicating an age of at least a few hundred years.

The many occurrences of large frost-crack polygons in the Abisko area seem to represent a whole spectrum of conditions. One extreme is fossil ice-wedges in former permafrost conditions in the eastern high mountains with wide, windswept valley bottoms above 1000 m altitude and thin snow cover. The other extreme is a small area of large frost-cracks on an alluvial fan at 730 m altitude in the western, snow-rich valley bottom of Kärkevagge. These crack patterns were discovered in September 1983 and their 1-metre deep wedges were excavated and interpreted as sub-fossil soil wedges (Bärring and Loman, 1983). After 30 years of intense studies, Kärkevagge valley still offers interesting surprises to the open-minded and keen observer.

Nivation hollows and other effects of nivation processes in Lappland

The term nivation was introduced by Matthes (1900) to indicate snow-patch erosion, resulting from intensified frost-weathering, slope wash and mass movements of debris near thick snow-drifts in mountain or tundra areas, where thick snow patches melt during summer or autumn each year. Thorn (1976) has made intensive studies on nivation processes. After field studies of two instrumented snow-patch sites in the Colorado Front Range — one on unconsolidated syenite debris, the other on gneiss bedrock — he arrived at the following conclusions. In contrast to a snow-free site, nivation increases the mechanical transport of sand, silt and clay by an order of magnitude. The snow-patch itself is protective, sediment removal being focused down-slope of the retreating snow margin. Chemical weathering is increased by a factor of 2 to 4 by a snow-patch. Mechanical weathering can be intensified due to increased freeze–thaw cycle effectiveness. In high latitudes, solifluction (gelifluction) is probably more important than in mid-latitude mountains. Its role in the nivation process can be observed in front of melting snow-patches on many sites in Lappland and other Arctic areas such as Iceland (Lewis, 1939; Rudberg, 1974).

In spite of the strong effects of snow drifting and the very regular location of late snow-patches — they remain and melt on exactly the same spots every summer — so far no clear examples of nivation hollows excavated into bedrock have been described from the Scandinavian mountains. Probably the 9000 years of post-glacial tundra conditions is too brief to produce clear nivation hollows in the common bedrock types of the Swedish mountains. Figure 3.5 shows a site of possible nivation hollows in the east-facing sides of the Nissunjokk canyon (Figure 3.2, rectangle 6 km south of Abisko). This canyon is about 3.5 km long, runs in a North-South direction, and has a depth of about 60-100 m below the surrounding plateaux of undulating mountain tundra. The bedrock is amphibolite, a dark, metamorphic rock which in this area is part of the Caledonian overthrust rocks. The canyon has been cut by the mountain stream, Nissunjokk, draining the beautiful trough valley, Nissunvagge. The valleys and the mountains, which reach elevations of 1700-1800 m (cf. Figure 3.2) have been completely covered by inland ice in the Weichselian and probably also earlier glaciations. The main glacier movement was from South to North through the valley, and then towards the North-West and West to the Atlantic coast near Narvik. Erratic boulders of granite, syenite and gneiss from the Precambrian basement rocks, South-east of the Caledonian mountain zone, are proofs of the ice flow from South-east to North-west. Similar canyons occur at many places in the area, particularly where glacial trough valleys of the mountain zone end, hanging above a somewhat over-deepened valley of higher order. This canyon was selected as a particular study site for nivation processes for several reasons. First, its dimensions are similar to canyon valleys in South Sweden, where fossil nivation hollows have been postulated recently (Rapp, 1982, 1983). Secondly, Nissunjokk's canyon is directed from North to South and acts as an efficient trap for drifting snow from predominantly westerly wind directions. Thirdly, it begins above the tree line with strong snow drifting, and continues into the birch forest and thus is an instructive site to study the differences of impact of snow drifting above and below the upper forest limit. Lastly, it is situated about 4 km from the Abisko research station, which makes it easily accessible for continuous studies in both winter and summer.

Figure 3.5 shows a nivation site of the east-facing side of the canyon. The large snowdrifts in early June demonstrate the effect of westerly snow drifting on the treeless plateau. The sides of the canyon are not a straight wall but a series of excavations, shaped like hourglasses. The upper part of the hourglass carries the snowpatch which, by 10 July, had melted to two small remnants. Below the 'waist' of the hourglass is a colluvial fan of mixed debris, deposited mainly by small snow avalanches from the nivation hollow to judge from the low slope angle of the fan — 23° to 28° — which is a much lower gradient than rockfall talus (above 35°). Other indicators of snow–avalanche transport and deposition is an even surface of the central part of the fan and low vegetation with scattered angular blocks and pebbles deposited on the vegetation. The stream has cut laterally into the lower part of the fan, with a slope of 42°, which reveals mixed material of blocks and fines in the fan, without sorting, stratification or rounding of particles.

The conclusions of these observations are that the sides of the canyon have steep, semicircular hollows cut in amphibolite bed-

Figure 3.5: Nivation hollows and colluvial fan, east-facing canyon wall of Nissunjokk stream south of Abisko, Sweden. Bedrock: amphibolite. Depth of canyon c. 70 m. (a) shows large, leeside snow drifts, 12 June 1983; (b) shows small snow-patch, 10 July 1983. Nivation and small snow avalanches have excavated part of the Weichselian till cover in pre-Weichselian nivation hollow in bedrock, and during 9000 years accumulated the colluvial fan, marked by avalanche erosion from above and stream-cutting at the base. Conclusion: The canyon and the nivation hollows in bedrock are probably pre-Weichselian in age and have both survived obliteration by Weichselian main ice. (Photos A. Rapp.)

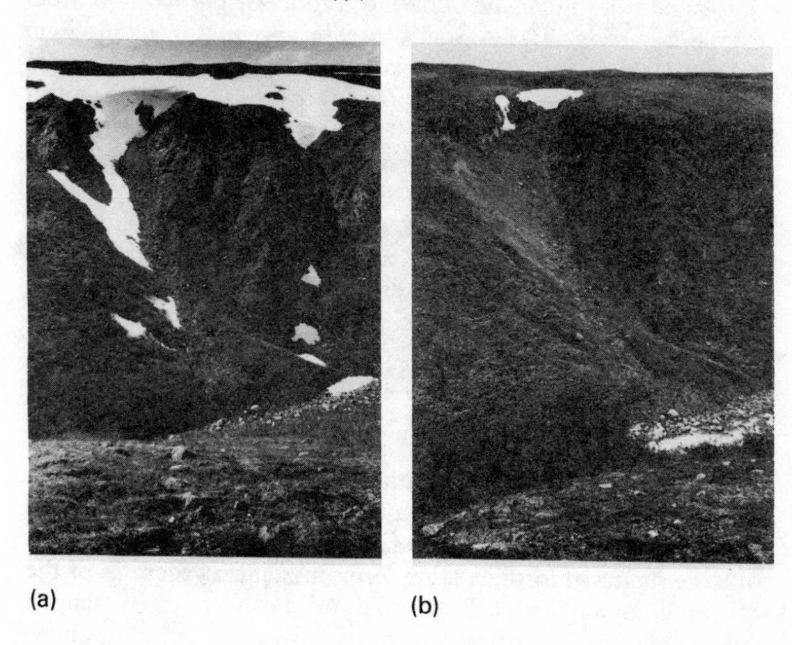

(a) (b)

rock. These bedrock hollows are thought to be pre-Weichselian nivation hollows, as they are now partly covered by Weichselian till, recognisable from the scattered erratic blocks of granite, syenite and gneiss in the till. The melting of the Weichselian inland ice occurred in this area about 9000 years BP. Consequently the postglacial nivation processes operating every spring and summer at this site have only partly cleaned the bedrock hollow from its till cover and — mainly by repeated small snow slides or avalanches — built the colluvial fan at the side of the stream. It is remarkable that snow-avalanche removal and deposition can create distinct deposits on slopes of only 50-70 m drop in elevation. Both the canyon and the nivation hollows existed before the Weichselian

Figure 3.6: Vertical section of an ice-wedge cast on the Laholm plain, South Sweden (Figure 3.7). The ground surface at the time of polygon formation is covered by wind-blown sand. The carpenter's rule is 1 m long. (Photo H. Svensson (1973).)

glaciation and survived the action of glacial scour. The studies of this and similar sites will continue to test the hypothesis of the mechanism of nivation processes, and for continued comparison with the supposed fossil nivation forms in South Sweden.

Reconstruction of former tundra environments in South Sweden

For South Sweden, Berglund (1979) summarised the climatic shifts after the Weichselian deglaciation as follows: 'the deglaciation pattern and the vegetation/soil changes indicate an important climatic amelioration slightly after 13 000 BP interrupted by a slight deterioration 12 000 to 11 800 BP, corresponding to Older Dryas, and a severe, long period of deterioration 11 000 to 10 200 BP corresponding to Younger Dryas.' These periods had a very severe tundra climate in Scania. The growing number of observations of ice-wedge casts found in gravel pits on the sandy plains of Scania and South Halland (Figures 3.6 and 3.7) prove severe permafrost conditions probably during the Older and

Figure 3.7: Fossil ice-wedge polygons as crop marks on the plain of Laholm, South-west Sweden. (From H. Svensson (1973).)

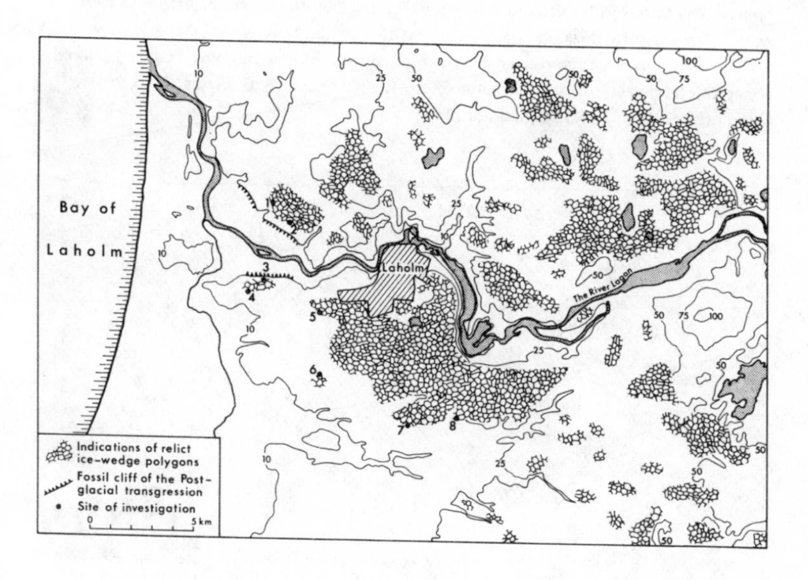

Younger Dryas as well as intervening phases of tundra climates in Scania (Svensson, 1973; Johnsson, 1982).

On the Laholm plain in South Halland near the south-western coast of Sweden (Figure 3.7), fossil ice-wedge polygons appear as crop marks in cultivated areas, and can be mapped by photography from light aircraft during dry periods in summer (Svensson, 1973). The stratigraphical characteristics of the polygon lines were studied in test pits, which revealed clear ice-wedge casts (Figure 3.8). The area is now densely cultivated and one of Sweden's main agricultural provinces, with wheat, rye, sugar beet and even maize grown successfully. The mean annual air temperature is 7.3°C and the mean annual precipitation 735 mm. The area is situated below the marine limit of the Weichselian glaciation. The former highest coast line is about 55-60 m asl. Glaciofluvial deposits, mostly of sand, dominate the plain area, which is the larger delta of the former River Lagan during deglaciation and subsequently, when isostatic land uplift caused a regression of the coast line. From an analysis of the distribution of polygons in relation to former coastlines and height above the present sea level, Svensson (1973) concludes that the polygons of the eastern, higher parts of the plains

Figure 3.8: Map of South Sweden with periglacial wind-polished boulders or bedrock. Key: 1. Fault scarp in Precambrian gneiss. 2. Locality with boulders or bedrock polished by winds in tundra periods after Weichselian deglaciation. Arrow shows easterly or westerly directions. 3. Locality with nivation hollows in bedrock. N = distinct forms; n = vague forms. 4. Town. (Based on Mattsson (1957) with additional observations by H. Svensson, G. Johnsson, T. Ahrnström, P. Schlyter, J. Åkerman (wind erosion) and A. Rapp (nivation).)

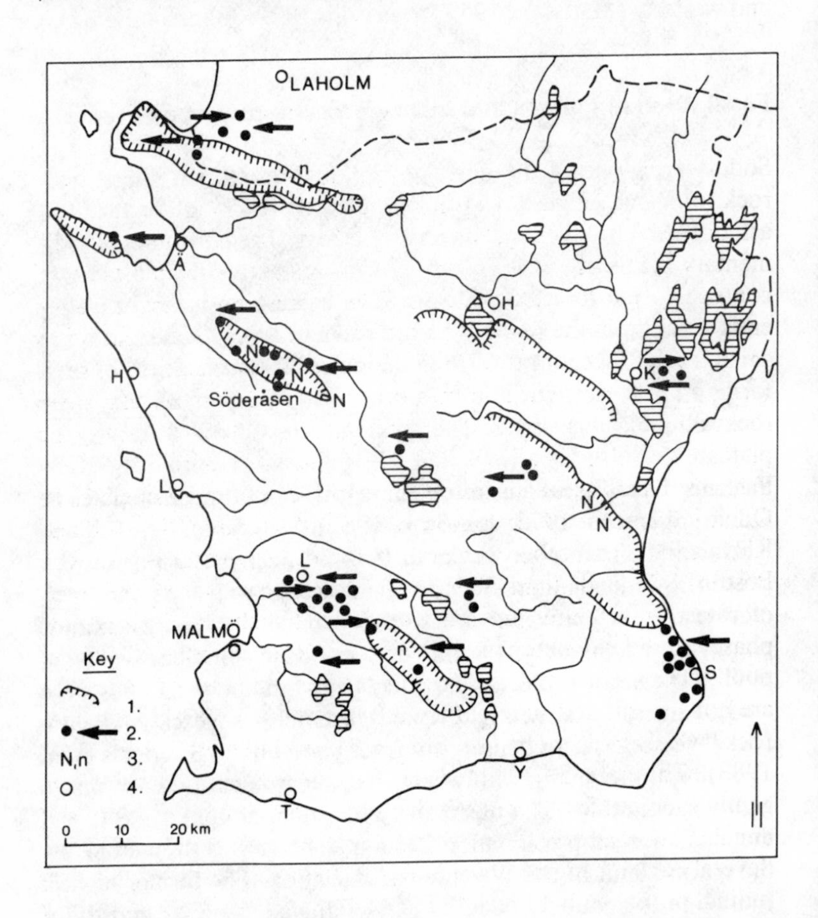

may have begun to form during the Oldest Dryas period, and that the youngest (lowest) polygons observed most probably were formed in the Younger Dryas period, ending about 10 200 BP. Similar, although less widespread, occurrences of fossil ice-wedge casts of several metres depth, and corresponding ice-wedge polygon patterns on the ground surface, have been observed at a large number of localities with sandy soils in South Sweden (Svensson, 1973).

Another sign of severe tundra climate in phases after the major deglaciation in South Sweden are occurrences of wind-polished bedrock and boulder pavements (Figure 3.8). Their wind fluting and facets of wind polishing are from two main directions: easterly and westerly (Mattsson, 1957; Svensson, 1972; Johnsson, 1982).

Fossil nivation hollows and other effects of nivation in Scania

Söderåsen is one of the horst blocks of Precambrian gneiss bedrock in Scania (Figure 3.9). It is a plateau reaching about 200 m asl, dissected by steep-sided canyon valleys. The surrounding sedimentary plains are at 40-60 m asl with mainly cultivated land, in contrast to the forested plateaux. The easternmost canyon valley ends abruptly in the South with the basin of Lake Odensjön. It is a small, round lake, only 160 m in diameter, but quite deep, 21 m. It forms an over-deepened basin, surrounded by a semicircular, steep rockwall and talus slopes. The lake surface is about 60 m asl, the plateau above the headwall is at 95-100 m asl (Figure 3.9). Many theories have been suggested to explain the genesis of Lake Odensjön and the other canyon valleys in Söderåsen, Skäralid and Klövahallar. The most widespread theory, suggested by L. von Post in 1938, is that the lake basin and the canyons were the result of catastrophic water erosion from glacial meltwater in one late phase of the Weichselian deglaciation — a 'superflood plunge-pool' theory. However, the rims of the canyon sides of Söderåsen are not straight and simple; they contain many hollows in the bedrock, which look more like fossil nivation hollows (Rapp, 1982, 1983). They are 20 150 m wide, have a semicircular back rim, gently sloping floors without incision by a stream trench, and angular rock debris over the bedrock on the floor and sides. Several have a hanging position to the main valley; that is, a longitudinal profile with a break in inclination between a gently sloping floor of the hollow and a steeper lower slope.

Another fact, inconsistent with the glaciofluvial plunge-pool theory, is that rounded, water-transported and sorted gravel, pebbles or boulders are missing in deposits corresponding to the hollows. Some of these deposits seem to be similar to the colluvial fans in Figure 3.5, and similar forms in actual nivation hollows in

Figure 3.9: Generalised map of Söderåsen, Scania, a gneiss plateau with canyon valleys and periglacial nivation hollows and glacial cirques. Key 1: Sandy plains with many ice-wedge casts and wind-polished rock surfaces. (Svensson, 1975; Johnsson, 1982) from periglacial tundra periods with permafrost and strong snow-drifting. 2. Main fault scarp slopes of Söderåsen. 3. Canyon valley with straight sides and a small stream. 4. Nivation hollows and cirques in bedrock of canyon sides. Nivation forms reflect strong snow-drifting from the west and east quadrants during tundra periods. 5. Peat land. 6. Elevation above sea. 7. Lake and stream. 8. Locality with wind-polished bedrock. Direction from E 10°S. (B. Ringberg, personal communication.) Arrow (right) marks direction of main Weichselian ice from the North-east.

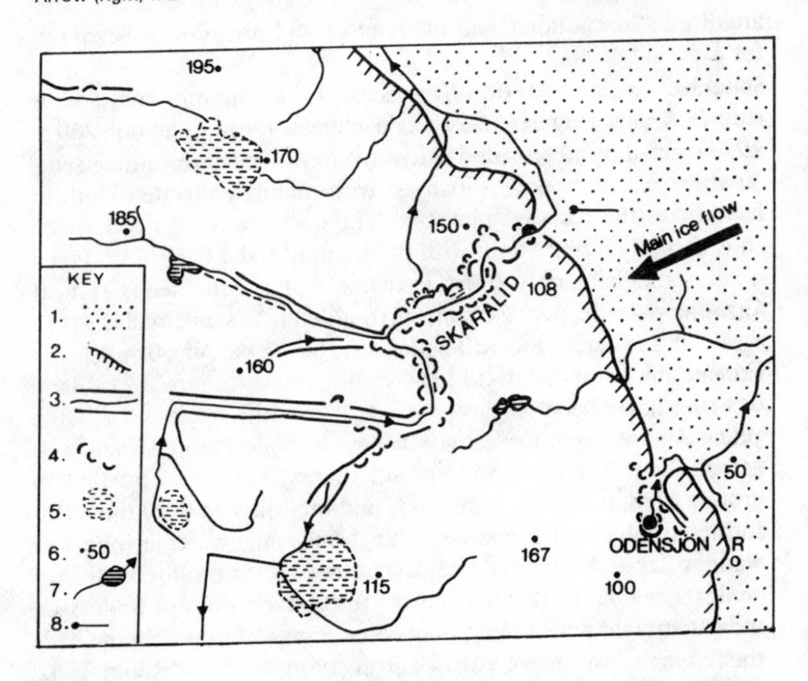

Söderåsen's canyons are partly covered by Weichselian till with scattered, large, rounded, erratic boulders on the surface. Hence the bedrock hollows are older than the Weichselian glaciation. In front of two of the larger cirque-like concavities of 200-400 m width in the sides of the Skäralid canyon are moraine ridges interpreted as local end-moraines (Rapp, 1983). It seems they were deposited in front of small glaciers, which probably existed due to strong snow drifting across the plateau of the horst ridge in the tundra periods of Older or Younger Dryas times after the Weichselian deglaciation.

If local glaciers could grow and be active at this low level and so far to the South in Sweden during short periods — a few hundred years in duration — of severe tundra climate after the Weichselian deglaciation, then local glaciers could have existed and created cirques during the much longer tundra periods in South Sweden before the Weichselian ice advanced over the area. It is supposed, based on C 14 datings of pre-glacial peat and other evidence that 40 000-50 000 years of tundra climate was predominant in South Sweden during the earlier phases of the Weichselian period (Berglund and Lagerlund, 1981). The Weichselian icesheet covered the area from about 21 000 BP to about 13 500 years BP; that is, only for a period of about 7000-8000 years. Re-growth and activity of small cirque glaciers have been reported from particular case studies at sites near sea level in western Norway (Mangerud *et al.*, 1979) and in the Lake District of England (Sissons, 1980), in the Younger Dryas period, and may be parallels to the cases postulated here from South Sweden.

Conclusions

Frost-crack polygons of different types have been discovered and described in the mountains of North Sweden during the last 20 years. So far only a few of these localities have been closely investigated. They seem to represent a wide spectrum of climatic and ground conditions, from sub-fossil ice-wedge casts formed during former permafrost conditions a few hundred years ago, to soil-wedges formed during non-permafrost conditions. Fossil ice-wedge casts in South Sweden are of larger depth and width, and indicate much more severe permafrost climate during phases after the Weichselian deglaciation in South Sweden such as the Older and Younger Dryas periods than the present tundra climate of the coldest parts of Lappland. The fossil indicators of strong wind action in South Sweden — wind-polished stones and bedrock, wind flutings — are much better developed than the similar forms in the coldest parts of Lappland. The predominant rock-polishing winds in South Sweden were of easterly directions, with a secondary maximum, more weakly developed, from the westerlies.

The processes of nivation (snow-patch erosion) are active now in the tundra mountains of Lappland. But clear nivation hollows have not been previously described, possibly for two reasons: (1) almost all good leeside sites have grown into glacial cirques over the last million years; and (2) the post-Weichselian time is, so far, too short for nivation hollows to form in Lappland. Recently, observations of possible pre-Weichselian nivation hollows in the side-walls of a canyon in a tundra area near Abisko, Lappland, have been made and may fit the theory of nival erosion and glacial protection. A similar genesis of long-lasting tundra conditions with permafrost, strong snow drifting, snow-patch erosion and pre-Weichselian creation of nivation hollows is likely for canyon valleys in South Sweden. Some of the hollows seem to have grown into small glacial cirques, 200-400 m wide and 50-100 m high, mainly during pre-glacial tundra periods, but also to a minor extent been rejuvenated by short-lasting, local glaciers, soon after the Weichselian deglaciation to judge from a few cases of local moraine-like ridges.

References

Bärring, L. and Loman, G. (1983), 'Jordkilar i Kärkevagge. (Soil wedges in Kärkevagge)', *Svensk Geogr. Årsb., 59.*

Berglund, B. (1979), 'The deglaciation of southern Sweden 13 500-10 000 BP', *Boreas, 8,* 89-117.

Berglund, B. and Lagerlund, E. (1981), 'Eemian and Weichselian stratigraphy in south Sweden', *Boreas, 10.*

French, H.M. (1976), *The Periglacial Environment* (Longmans, London).

Friedman, J.O., Johansson, C.-E., Oskarsson, N., Svensson, H., Thorarinsson, S. and Williams, R.S. (1971), 'Observations on Icelandic polygon surfaces and palsa areas, photo-interpretation and field studies', *Geogr. Ann., 53A,* 115-45.

Johnsson, G. (1982), 'Periglacial vindslipning och torbildning på Höörsandstenen', *Svensk Geogr. Årsb., 58.*

Lewis, W.V. (1939), 'Snow patch erosion in Iceland', *Geogr. J., 94,* 153-61.

Lundqvist, J. (1962), 'Patterned ground and related frost phenomena in Sweden', *Sver. Geol. Unders. Årsb., 55.*

Mangerud, J., Larsen, E., Longva, O. and Sønstegaard, E. (1979), 'Glacial history of Western Norway 15 000-10 000 BP', *Boreas, 8,* 179-87.

Matthes, F.E. (1900), 'Glacial sculpture of the Bighorn Mountains, Wyoming', *US Geol. Surv. 21st Ann. Rep., 2,* 167-90.

Mattsson, Å. (1957), 'Windgeschliffenes Gestein im südlichsten Schweden und auf Bornholm', *Svensk Geogr. Årsb., 33.*

Mattson, Å., (1962), 'Morphologische Studien in Südschweden und auf Bornholm über die nichtglaziale Formenwelt der Felsenskulptur', *Medd. Lunds Univ. Geogr. Inst. Ser. Avh., 39.*

Melander, O. (1977), *Geomorphological Map 30 I Abisko, etc. SNV PM 857,* Stockholm.

Péwé, T.P. (1964), 'Ice-wedges in Alaska', *Geol. Soc. Am. Spec. Pap., 76.*

Rapp, A. (1982), 'Periglacial nivation cirques and local glaciations in the rock canyons of Söderåsen, Scania, south Sweden', *Dansk Geogr. Tidsskrift, 82.*

Rapp, A. (1983), 'Impact of nivation on steep slopes in Lappland and Scania,

Sweden', *Abh. Akad. Göttingen, 35.*

Rapp, A. and Annersten, L. (1969), 'Permafrost and tundra polygons in northern Sweden', in Péwé, T.L. (ed.), *The Periglacial Environment — Past and Present*, (McGill-Queens University Press, Montreal), pp. 65-91.

Rapp, A. and Clark, M. (1971), 'Large non-sorted polygons in Padjelanta National Park, Swedish Lappland', *Geogr. Ann., 53A*, 71-85.

Rudberg, S. (1974), 'Some observations concerning nivation and snowmelt in Swedish Lappland', *Abh. Akad., Göttingen, 29.*

Semmel, A. (1969), 'Verwitterungs und Abtragungs erscheinungen in rezenten Periglacialgebieten (Lappland and Spitsbergen)', *Würzburger Geogr. Arb., 26.*

Sissons, J.B. (1980), 'The Loch Lomond advance in the Lake District, northern England', *Trans. R. Soc. Edin.: Earth Sci., 71*, 13-27.

Svensson, H. (1973), 'Distribution and chronology of relict polygon patterns on the Laholm plain, the Swedish west coast', *Geogr. Ann., 54A*

Thorn, C.E. (1976), 'Quantitative evaluation of nivation in the Colorado Front Range', *Bull. Geol. Soc. Am., 87*, 1169-78.

Vilborg, L. (1977), 'The cirques of Swedish Lappland', *Geogr. Ann., 59A*, 89-150.

von Post, L., (1938): Odensjön, Skäralid, Klövahallar. Sv. Turistfőren. små häften, Skåne 1. Stockholm.

Washburn, A.L. (1979), *Geocryology* (Arnold, London).

4 GEOMORPHOLOGICAL DEVELOPMENT OF MODERN COASTLINES: A REVIEW

David Hopley

Due to the amount of water locked up in glacier ice, sea level stood more than 100 m below its present position 18 000 year ago. Subsequently, all continental shelves have experienced rapid drowning which apparently denies the general applicability of schemes differentiating between submergence and emergence (Johnson, 1919). However, sea-level behaviour relative to the land is not globally uniform, even in tectonically stable areas. This is due to a redistribution of ice load from the glaciated northern hemisphere continents (glacio-isostatic unloading) to a water load over the ocean basins (hydro-isostatic loading) during deglaciation. Therefore, the time that sea level has been at, or close to, its present position varies from mere hundreds of years to more than 6000 years.

Significant morphological differences might be expected between shore zones, due to differences in the time processes that have operated in the present-day shore zone. The present chapter also considers why coastal landform development in some areas is much more advanced than might be expected.

Geographical variation in the time of achievement of modern sea level

Bloom (1967) revived some older speculation on coastal isostatic deformation by the post-glacial rise of sea level. He suggested that the load of water added to the continental margins by the transgression would produce subsidence of coastal areas in proportion to the average depth of water in the vicinity. Subsequently, Walcott (1972) showed how relative sea level at great distances from the glaciated regions was affected by the meltwater loading of the ocean floor. Chappell's (1974) work concentrated on the critical hinge zone between oceans and continents. He suggested an average depression of ocean basins of about 8 m and mean upward

Figure 4.1: A variety of sea level curves from different parts of the world related to the zones of Clark *et al.* (1978), and showing the contrasting periods of time since the first achievement of present sea level. Although not shown in the references, acknowledgement is made to the authors of the various sea level curves.

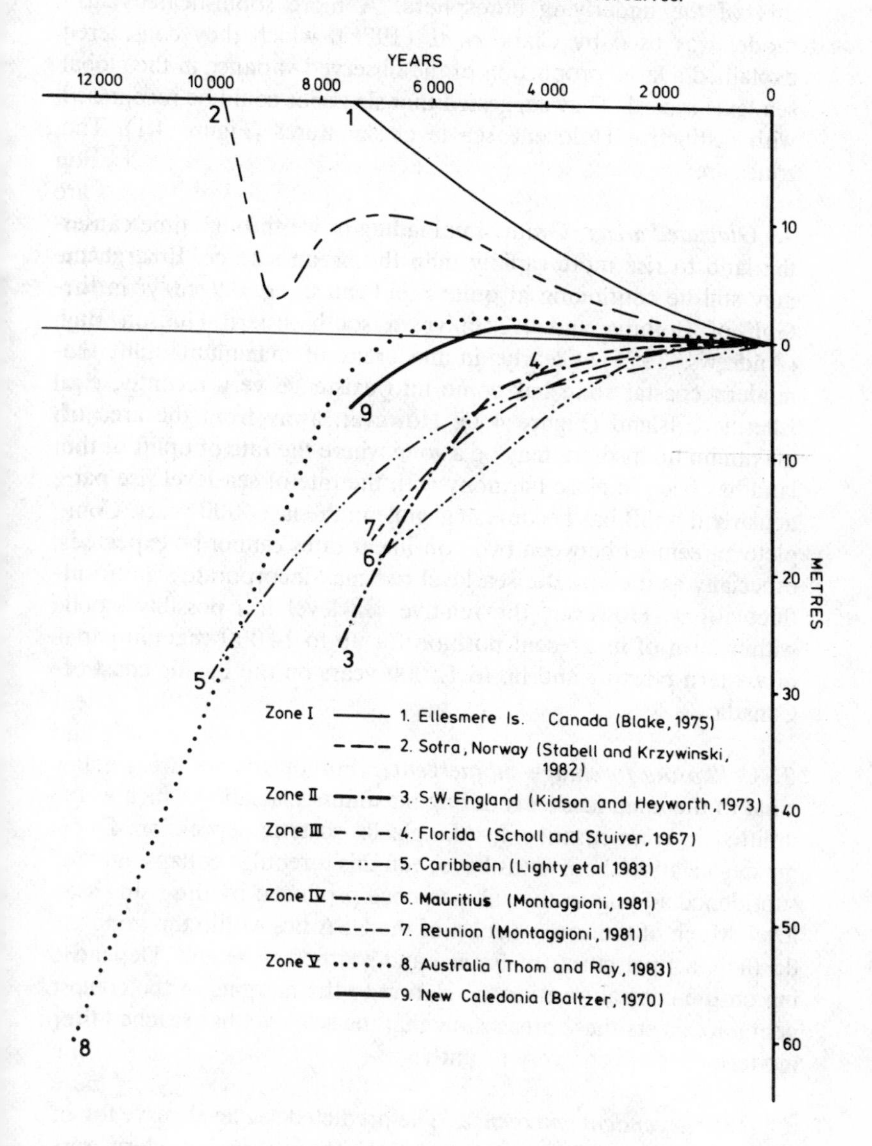

movement of continents of about 16 m relative to the centre of the earth in the last 7000 years. Deflection on the ocean margin hinge zone was shown to vary with continental shelf geometry and rigidity of the underlying lithosphere. A more sophisticated earth model was used by Clark *et al.* (1978), which they considered explained a large proportion of the observed variance in the global sea level record. They suggested that six zones could be recognised with distinctive Holocene sea level signatures (Figure 4.1). The zones are:

1. *Glaciated areas.* Gradual unloading of ice through time causes the land to rise more rapidly than the ocean surface. Emergence may still be continuing at quite rapid rates, e.g. 0.9 cm/yr in the Gulf of Bothnia and 1.2 cm/yr in south-eastern Hudson Bay (Andrews, 1975). Clearly, in the areas of maximum uplift the modern coastal zone has come into existence very recently, e.g. Ellesmere Island (Figure 4.1). However, away from the area of maximum uplift there may be a zone where the rate of uplift of the land has been in close harmony with the rate of sea-level rise particularly if uplift has become negligible in the last 6000 years. Complete agreement between two non-linear rates cannot be expected, especially as the eustatic sea-level rise may incorporate significant fluctuations. However, the relative sea level has possibly stood within 10 m of its present position for up to 14 000 years in parts of western Norway and up to 12 000 years on the Pacific coast of Canada.

2. *Collapsing forebulge submergence.* Forebulges are areas marginal to the main icesheets during maximum glaciation which were uplifted in compensation for the glacio-isostatic depression. During deglaciation these areas have seen this forebulge collapsing, the subsidence adding to the submergence produced by the rising sea level. Much of the eastern coast of the USA lies within this zone as do the Channel coasts of France and southern England. Depending on the coastal location in relation to the margins of the continental icesheets these areas show that the sea level has reached the modern coastal zone very recently.

3. *Time-dependent emergence.* The predicted sea level curve for a small proportion of the northern hemisphere shows modern sea level essentially being achieved approximately 3000 years BP.

4. *Oceanic submergence.* Much of the intertropical zone lies in an area where continued submergence is predicted. Data from Micronesia in the Pacific, and Mauritius and Reunion in the Indian Ocean, seem to conform, showing the modern coastal zone inundated only very recently.

5. *Oceanic emergence.* Most of the southern oceans are predicted to have achieved modern sea level as early as 6000 years BP with some degree of subsequent emergence up to about 2 m at 5000 years BP. Data for Australia, New Caledonia, New Zealand, atolls of the southern Pacific and parts of South America illustrate the comparatively long period of modern sea level and thus of modern coastal evolution.

6. *Continental shorelines.* As the continents are all being isostatically uplifted, dependent upon the position of the critical hinge line, all continental shorelines have the potential for emergence. This may merely produce an earlier achievement of modern sea level than is predicted for the region. In Australia, for example, modern sea level may have been achieved shortly after 7000 years BP (Hopley, 1983). Elsewhere it may also enhance the original emergence pattern to produce emergence of several metres.

Taking into account the fact that the modern coastal zone lies within a vertical band of several metres, determined by variations in annual mean sea level, tides and wave action, both the mathematical models and the regional evidence available clearly show that activation of the present littoral zone has taken place in as little as 100 years to a time-span as great as 7000 years. If the vertical range of the coastal zone is made as great as 10 m then, in some restricted areas of tectonic or isostatic uplift, the present coastal zone may have been evolving for as much as 14 000 years.

Rates of coastal processes

The establishment of rates of coastal processes can be made on three scales: (i) short-term measurements, for example, of beach changes by dumpy level survey, or similar short-term surveys of rocky shore erosion using micro-erosion meters; (ii) surveys of changes in historic times using photographs, maps and charts. Such changes may be established over hundreds of years; and (iii) long-term changes based on geological evidence and using radiometric dating techniques. Such methods may produce complementary results, although this appears to be rare. For example, studies of vertical coral reef accretion commonly use two timescale-contrasted methods. Short-term modern rates are derived from the alkalinity depletion of seawater passing over the reef flat, and actual rates derived vary according to reef zone from 0.4 to > 7 mm/yr. Longer-term rates, though the Holocene, may be made using radiocarbon-dated reef cores. A very wide range of growth rates have been quoted from such methods but many figures incorporate the deposition of detrital materials (Davies and Hopley, 1983). When *in situ* reef framework alone is considered a highly compatible value of between 7 and 8 mm/yr is achieved. However, in general the extrapolation of short-term rates to a geological timescale is not advised. Deposition and erosion may not occur at rates which are uniform through time. High-energy, low-frequency events, such as storms, produce an irregularity of process rates in many environments, including coastlines.

Short-term limestone erosion rates obtained by Trudgill (1976a and b, 1979) from the emerged atoll of Aldabra in the Indian Ocean in particular have been extended over the full period of the last glacial period (*c.* 100 000 years) by several authors. However, as Trudgill himself pointed out, extrapolation of this type is speculative as it assumes that the climate of a two-year period of measurement can be equated with past climates. Erosion rates obtained in supratidal areas may also be accelerated at the present time by salt spray or, at the very least, by the salt-laden atmosphere. In addition, blue-green algae, with an ability to bore into limestone surfaces, may be found up to 10 m above high tide level. If sea level fell by more than 20 m these areas from which micro-erosion rates may be obtained today would be isolated from both salt spray and the boring activities of the algae. The erosion rates could not be sustained through the long period of last glacial low sea level.

Extension of geological time-rates to areas other than that from which they are obtained must also be speculative for they may include the effects of environmental changes which are of local significance. Changes in climate in particular can affect the coastal zone. Periglacial shore platforms are formed primarily by frost action assisted by waves, a process which leads to very rapid platform development. It has been suggested that similar erosion took place in the British Isles during numerous phases of Pleistocene cold climate. Extensive raised or submerged platforms need not necessarily be considered as having formed during long interglacial periods of relative sea level stability.

In spite of these problems, rates which are applicable to the period of Holocene coastal evolution may be attempted. Depositional rates are very dependent on local sediment yield. However, Thom *et al.* (1981), in a study of beach barrier progradation in New South Wales, suggest two phases subsequent to sea level stabilisation (which here took place prior to 6000 years BP). Up to 4000 BP progradation was rapid, largely from sand sources moved landwards across the continental shelf by the post-glacial transgression. These early sand sources almost certainly include material reworked from Pleistocene interglacial barriers. Subsequently, the rate of progradation has slowed, and may even include erosional episodes as further deposition is dependent on local fluvial sediment yield or longshore drifting. However, data from the Shoalhaven delta, 120 km south of Sydney, suggest that a third phase of accelerating progradation may occur where intitial rapid barrier progradation across an embayment encloses a coastal lagoon into which a major river flows. Once the shelf sand source is depleted, negligible progradation may result until the river has infilled the lagoon as a deltaic plain. Subsequently, fluvial sand may be delivered to the coast, initiating a new phase of barrier progradation. Thus, whilst the absolute rate of progradation may depend on local sediment yield, the relative rate can be related to the length of time that sea level has been stable close to its present position.

Width of modern shore platform cutting is similarly dependent on this period of time. A wide range of figures are quoted in the literature, apparently dependent on rock type, wave energy, subaerial climate and local bioerosional activity. For hard rock coastlines, most lie within the range of 0 to 10 mm/yr. Trudgill's figures for horizontal notch cutting in Aldabra limestone ranged from 1 to 7 mm/yr, depending on exposure. A figure of 5 mm/yr would produce a platform of 30 m width in an area in which sea level had been constant for 6000 years but less than 5 m width where modern relative sea level has been achieved in the last 1000 years. This presumes that all other factors are similar, a situation which can be rarely achieved in the natural environment. A further complicating factor is the probable deceleration of platform development with increasing width of the platform and cliff collapse subsequent to undercutting. None the less, distinct differences should exist between, for example, the rocky coasts of Australia (more than 6000 years of Holocene development) and much of the coast of the USA or the English Channel coasts of England and France (less than 1000 years).

Two contrasting examples of coastal development dependent on Holocene sea levels

Two Australian examples of coastal geomorphology may be compared with northern hemisphere equivalents to indicate that the length of time that the sea has stood in its present position relative to the land is an important factor in coastline development.

Coastal barriers in New South Wales and Delaware (Figures 4.2 and 4.3)

Coastal barriers in New South Wales have been prograding for over 6000 years. Details are available for a number of them (see Thom *et al.*, 1981). That at Moruya shows a number of characteristic features. At the base of the section, at depths of more than 30 m, is an estuarine clay and organic mud of early Holocene age, probably formed behind a landward migrating barrier during the post-glacial transgression. They are overlain by a transgressive shelly sand and gravel facies. Subsequent to the stabilisation of sea level, a regressive facies of nearshore shelly sand overlain by beach ridges has prograded from the most landward position reached by the shoreline, the rate of progradation declining with time. More than two-thirds of barrier progradation took place between 6500 and 5000 years BP.

The east coast of the USA also has numerous barrier systems. In New Jersey, Delaware, Maryland and Virginia the coast is deeply embayed but fringed with long transgressive barriers which enclose numerous lagoons with large tracts of saltmarsh. The Delaware barriers have been examined in detail by Kraft and colleagues (e.g. Kraft, 1971; Kraft *et al.*, 1979). Sea level has been rising right up to the present in this area producing a typical receded (landward-migrating) barrier. The system includes thin, narrow, coast-parallel dunes, a beach system, washover fans, flood tidal deltas and nearshore marine offshore bars. A nearshore marine

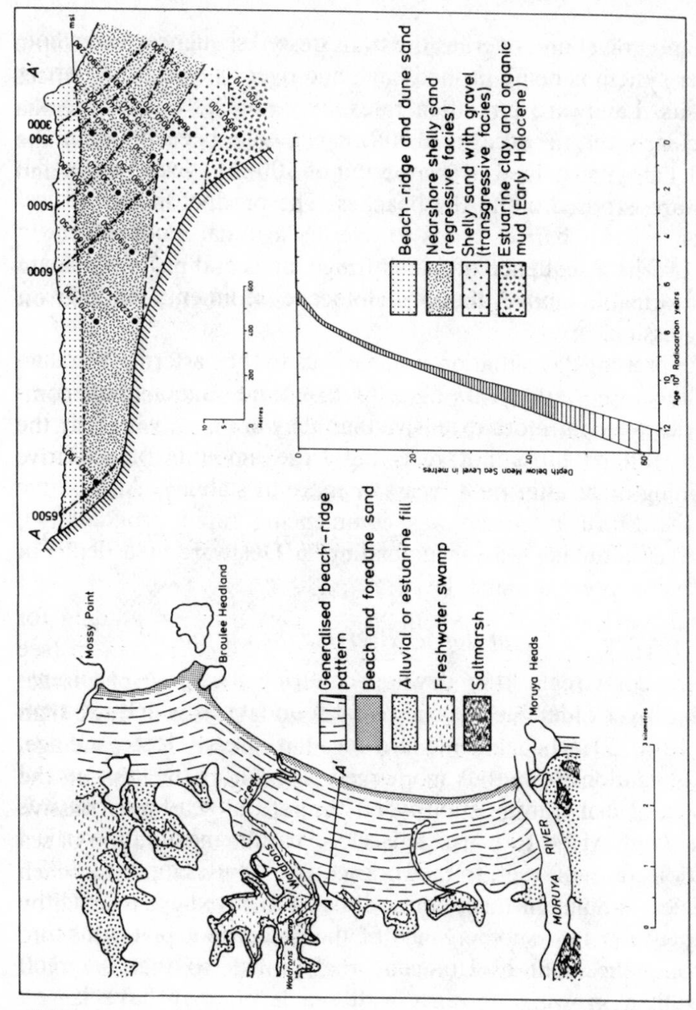

Figure 4.2: Moruya barrier system, NSW, Australia with local sea level curve (after Thom *et al.*, 1981 and Thom and Roy, in Hopley, 1983).

sand and gravel unit overlies the transgressed sequence. The whole barrier system is being pushed back and over its own back-barrier deposits. Landward migration rates, expressed as coastal erosion particularly during occasional hurricanes have been between 0.4 and 3.1 m/yr sustained over a period of 100-200 years. Saltmarsh peats are exposed along the beaches. The product in the vertical sequence is for barrier sands to overlie lagoonal muds which, in turn, overlie a sequence of marsh fringe muds and peats which are unconformable above the pre-Holocene sediments undergoing transgression.

The present-day situation is analogous to the eastern Australian coastline over 6500 years ago. At that time, lagoons and open estuaries were far more extensive than they are now, with only the largest and/or those not receiving large amounts of sediment remaining open after 6000 years or more of stability. At the time that the Moruya barrier was commencing rapid progradation, intertidal saltmarsh was accumulating in Delaware at a depth of 15 m below present marsh level.

Coral reefs of the Great Barrier Reef and the Caribbean

Modern coral reefs have developed during the Holocene transgression over older Pleistocene reefal foundations which are typically 10 to 20 m beneath modern reef-flat levels (Hopley, 1982). The foundations were thus submerged only during the later part of the post-glacial transgression when the rate of sea-level rise was about 7 mm/yr, a rate which is close to the maximum vertical accretion rate of reef framework (Davies and Hopley, 1983). However, in many instances there appears to have been a lag in the time between first submergence of the antecedent platforms and the commencement of Holocene reef growth so that the reef, rather than growing upwards with sea level, may have lagged behind by several metres.

The Great Barrier Reef has developed in response to the sea-level rise seen in Figure 4.4, although some regional variation may have existed. Fringing reefs of the inner shelf, in particular, may exhibit an emerged reef of almost 1.5 m elevation. As sea level came close to stabilisation 6500 years ago, many reefs, lagging behind the rise by only a few metres, reached sea level by 5000 BP, some as early as 6000 BP (Hopley, 1982). Subsequently, most reef growth has been in the horizontal plane, developing extensive reef flats over which a distinctive morphological zonation has had time

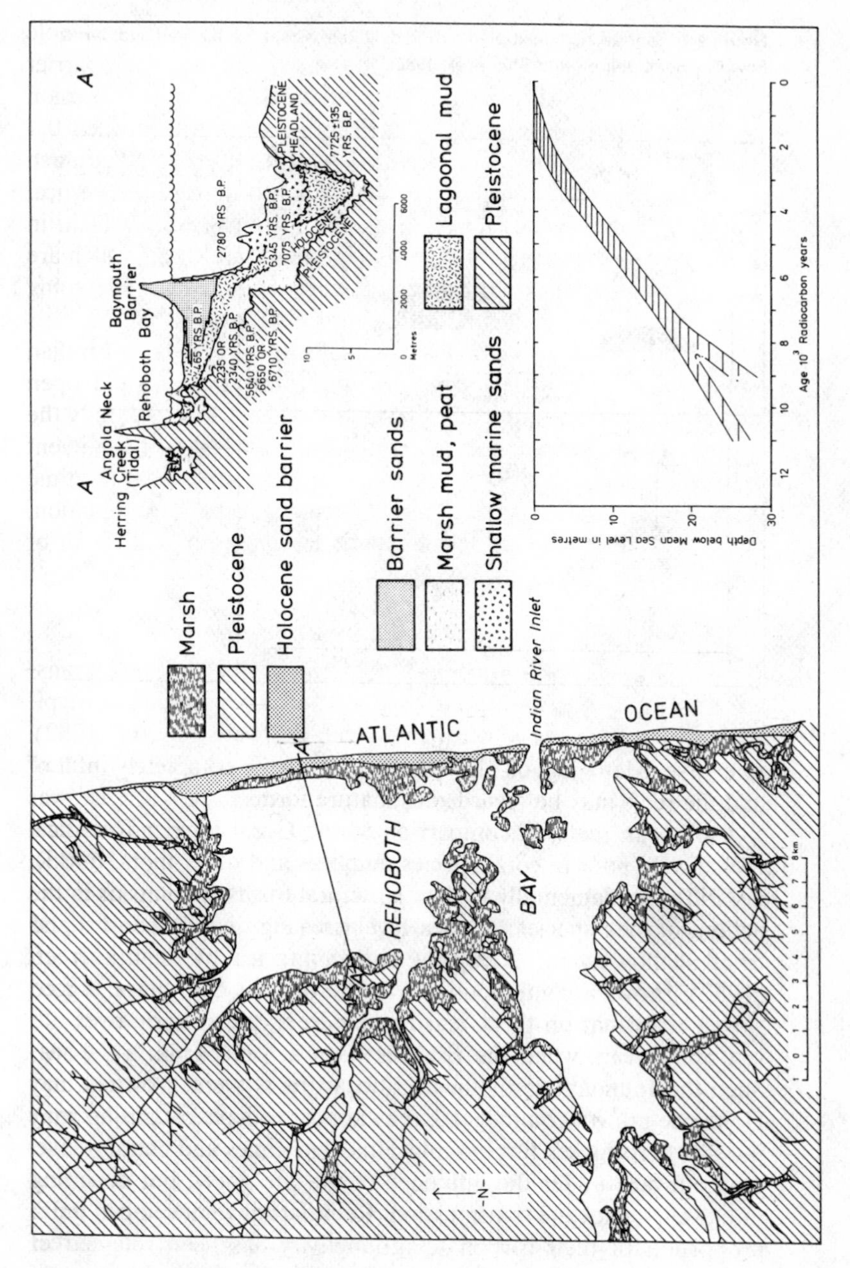

Figure 4.3: Rehoboth bay barrier system, Delaware, USA with local sea level curve (after Kraft, 1971; Kraft *et al.*, 1979).

Figure 4.4: Contrasting age structure of Redbill Reef, Great Barrier Reef and Carrie Bow Cay Reef, Belize (after Shin *et al.*, 1982).

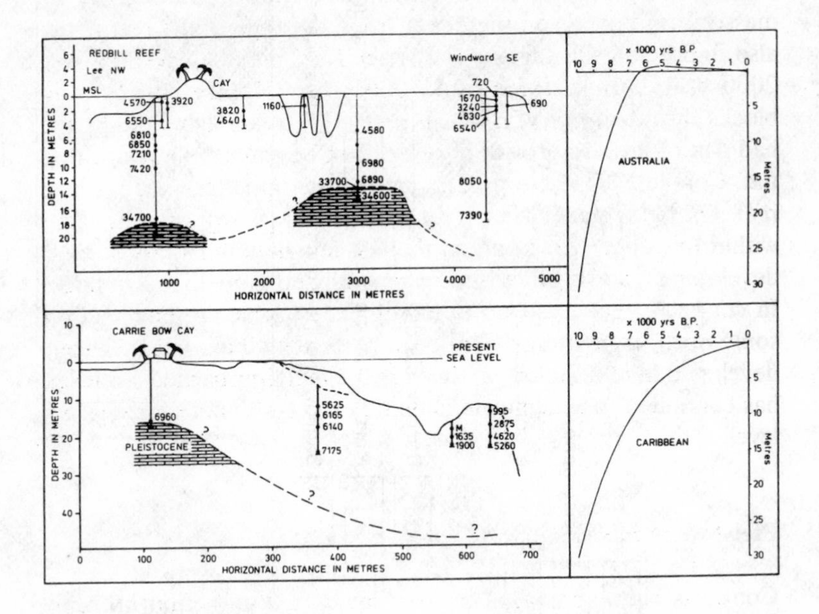

to evolve. Many lagoons have been partially or completely infilled and the reefs may be regarded as mature forms.

Caribbean reefs, in comparison to the Great Barrier Reef, are biologically poor in coral species numbers and other biota, but do not differ fundamentally in their structural forms, habitats or interactions of their species. The sea-level rise (Figure 4.4), constructed from shallow water *Acropora palmata* facies by Lighty *et al.* (1982), shows a continuous rise to the present day, but at a rate slower than that on the Great Barrier Reef, particularly over the last 5000 years when the rise has been at only 1 mm/yr. Thus, once the foundations of the Caribbean reefs were drowned and Holocene growth commenced, they appear to have caught up with the slowly rising sea level in mid-Holocene times and continued to grow upward under the influence of the prolonged but declining rate of the Holocene transgression. Reef flats have developed but a proportion of their carbonate productivity has been retained in upward growth. Reef flats are therefore young, often coral-covered and with poorer morphological zonation than their Great

Barrier Reef counterparts. Reef blocks, massive slabs of coral broken from the reef front during cyclones, are far rarer on Caribbean reefs. Although this is partly the result of the scarcity of massive head corals on these reef fronts, the age of the reef flat is also important. On the Great Barrier Reef many blocks are over 2000 years in age, some over 4000 years (Hopley, 1982). Any blocks thrown up on Caribbean reef 4000 years ago would have had 4 m of upward growth burying them beneath the modern reef flat. Coralline algae are prolific only in the upper few metres of a reef. On the Great Barrier Reef the upward growth of reefs came within this upper zone only in the last few hundred years of their development and algal crusts are generally no more than 2 m thick. In contrast, algal crusts up to 8 m thickness have accumulated on some open, high-energy Caribbean reefs. The initial algal veneer, developed in mid-Holocene times as the reef approached sea level, has continued to accumulate, aided by the continuously rising sea level.

Pleistocene inheritance

Contrasts between coastal regions having a long period of stable sea level and those with a continuously rising sea level, such as described above, are not always as great as might be expected. The probable reason for this is the degree to which modern coastal landforms are inherited from earlier forms, particularly the last interglacial when the position of the sea against the land in stable areas was within a few metres of its present position. Older deposits have been reworked and incorporated into modern sediments (Figure 4.5) and old cliff lines and platforms have been rejuvenated.

Not all examples are as obvious as Figure 4.5. Quite often the incorporation of Pleistocene material is only picked out when, for example, a shell-hash date from a barrier sequence is anomalously old. In South Australia there are extensive shingle ridges along the eastern shoreline of upper Spencer Gulf which contain abundant remains of the estuarine cockle *Anadara trapezia* and are elevated 3-5 m above present sea level (see Hails *et al.*, in Hopley, 1983). A last interglacial age was assigned to these ridges as *A. trapezia* is a Pleistocene indicator. However, mixed shell samples yielded Holocene dates. The Holocene age is probably accurate but the abun-

Figure 4.5: Raised (+6 m) shore platform and boulder beach of last interglacial age overlain by periglacial head, Westward Ho, Devon. Boulders from the raised beach are clearly incorporated in the modern beach, which overlies a shore platform. Although this is extending into the higher platform it is possible that this too was initiated at an earlier period.

dance of *A. trapezia* implies substantial reworking of Pleistocene deposits. Whilst whole shells of *A. trapezia* may be clearly identified as contaminant, other species still common in the area, finer fragments and non-biogenic deposits cannot be so readily identified. This example, however, suggests that incorporation of Pleistocene deposits in Holocene beaches may be common.

Platform rejuvenation also may not be easy to interpret. Once within even the uppermost part of the wave zone, rates of lowering of 0.02-1.8 mm/yr (Gill and Lang, 1983) can reduce a Pleistocene platform by up to 9 m in 5000 years. At the present time, the last interglacial platforms, up to 5 m above present levels, are in a location on most coasts which puts them in an area of intensive weathering by constant wetting and drying. At least the surfaces of these platforms may also have been prepared by weathering during the last glacial. For example, along both sides of the English Channel remnants of last interglacial platforms and their related boulder beaches show severe frost-shattering from Devensian periglacial conditions and, in Singapore, a platform of similar age has been completely rotted to several metres depth by severe tropical weathering. Bloom (1978, Figure 19.4, p. 445) illustrates a wide shore platform cut in chalk on the Channel coast of France at Ault. Nearby, this same platform is overlain by last glacial periglacial deposits, including mammoth tusks and also by a Paleolithic site with Levallois stone tools. Clearly, the recent attainment of modern sea level in this area has helped to preserve this site, but the indication of the minimally modified nature of the modern shore platform is equally important.

Conclusion

Considerable advances have been made in the last 25 years in understanding geomorphological processes and the rates at which they currently operate. In addition, most landscapes are recognisably polygenetic, having evolved during the Quaternary when rapidly changing climates have determined the superimposition of process-determined landscape types. In the coastal zone, processes and rates of change have been measured and, although regional variations in coastal morphology related to climate are recognised (Davies, 1973), superimposition of landscape types has not been considered important. This is largely the result of Quaternary glacio-eustatic changes in sea level which have determined that the narrow zone in which the modern coast is located has been activated over relatively short periods (< 10 000 years) of presumably similar interglacial climates. These are interspersed with long (> 100 000 years) periods of isolation when glacial period sea levels stood below their present position somewhere on the continental shelf. Due to hydro-isostatic responses, however, considerable variation exists in the period of activation of the modern coastal zone with potential for regional morphological variation. Regional coastal morphology studies should recognise the time available for processes operating in the modern littoral zone and the rate at which coastal processes operate. Significant morphological differences should be expected between shore zones as the result of differences in the time of activation of the zone. However, in some areas where activation is relatively recent, development of coastal landforms appears clearly in advance of what might be expected in the time available. The degree of inheritance from Pleistocene landforms is a further factor producing variation in the stage of development of the modern coast. Morphology is thus a result of the interplay of: (i) the history of Holocene sea level change in combination with (ii) the nature and rate of modern processes superimposed over, and (iii) the Pleistocene littoral zone reoccupied during the Holocene transgression. For coastal management purposes, the history of coastal development in areas with several millenia of stable sea level may act as a guide to those areas in which stability has yet to be attained.

References

Andrews, J.T. (1975), *Glacial Systems: An Approach to Glaciers and their Environments* (Duxbury, Massachusetts).

Bloom, A.L. (1967), 'Pleistocene shorelines: a new test of isostasy', *Bull. Geol. Soc. Am., 78,* 1477-94.

Bloom, A.L. (1978), *Geomorphology: A Systematic Analysis of Late Cenozoic Landforms* (Prentice Hall, Englewood Cliffs).

Chappell, J. (1974) 'Late Quaternary glacio- and hydro-isostasy on a layered earth', *Quat. Res., 4,* 429-40.

Clark, J.A., Farrell, W.E. and Peltier, W.R. (1978), 'Global changes in postglacial sea level: a numerical calculation', *Quat. Res., 9,* 265-87.

Davies, J.L. (1973), *Geographical Variation in Coastal Development* (Oliver & Boyd, Edinburgh).

Davies, P.J. and Hopley, D. (1983), 'Growth facies and growth rates of Holocene reefs in the Great Barrier Reef', *BMR J. Austr. Geophys., 8,* 237-51.

Gill, E.D. and Lang, J.G. (1983), 'Micro-erosion meter measurements of rock wear on the Otway coast of southeast Australia', *Mar. Geol., 52,* 141-56.

Hopley, D. (1982), *Geomorphology of the Great Barrier Reef: Quaternary Development of Coral Reefs* (Wiley-Interscience, New York).

Hopley, D. (ed.) (1983), 'Australian sea levels in the last 15,000 years', *Dept. Geog., James Cook Univ., Monogr. Ser. Occ. Pap., 3* (Townsville).

Johnson, D.W. (1919), *Shore Processes and Shoreline Development* (John Wiley, New York).

Kraft, J.C. (1971), 'Sedimentary facies pattern and geologic history of a Holocene marine transgression', *Bull. Geol. Soc. Am., 82,* 2131-58.

Kraft, J.C., Allen, E.A., Belknap, D.F., John, C.J. and Maurmeyer, E.M. (1979), 'Processes and morphologic evolution of an estuarine and coastal barrier system', in Leatherman, S.P. (ed.), *Barrier Islands from the Gulf of St Lawrence to the Gulf of Mexico* (Academic Press, New York), 149-83.

Lighty, R.G., MacIntyre, I.G. and Stuckenrath, R. (1982), '*Acropora palmata* reef framework: a reliable indicator of sea level in the Western Atlantic for the past 10,000 years', *Coral Reefs, 1,* 125-30.

Thom, B.G., Bowman, G.M., Gillespie, R., Temple, R. and Barbetti, M. (1981), 'Radiocarbon dating of Holocene beach-ridge sequences in southeast Australia', *Dept. Geog., Fac. Military Stud., Univ. NSW, Duntroon, Monogr., 11.*

Trudgill, S.T. (1976a), 'The marine erosion of limestones on Aldabra Atoll, Indian Ocean', *Z. Geomorphol. Suppl., 26*, 164-200.
Trudgill, S.T. (1976b), 'The subaerial and subsoil erosion of limestones on Aldabra Atoll, Indian Ocean', *Z. Geomorphol. Suppl., 26*, 201-10.
Trudgill, S.T. (1979), 'Surface lowering and landform evolution on Aldabra', *Phil. Trans. Roy. Soc. Lond., B., 286*, 35-45.
Walcott, R.I. (1972), Past sea levels, eustasy and deformation of the earth', *Quat. Res.*, *2*, 1-14.

5 GEOMORPHOLOGICAL PROCESSES, SOIL STRUCTURE, AND ECOLOGY

A.C. Imeson

It is not just on coral reefs that landforms evolve in association with living organisms. Many other geomorphological processes are interwoven with ecological ones from which they can be differentiated only with difficulty. Equally, landforms evolve through the displacement of soil particles or as a result of processes that operate within the soil. The soil itself forms a medium over and through which transported material passes. Therefore, ecological and pedological conditions must be examined whenever geomorphological processes are considered with respect to the landforms on which they operate. Any isolation of geomorphological processes from the pedological and ecological environments in which they are observed portrays landform evolution incompletely.

In this chapter, linkages between geomorphological processes, soil structure and organisms are described. The need for considerations of sediment transport mechanisms on hill slopes to be accompanied by parallel investigations of pedological and biological factors is demonstrated. In the case of soil, in particular, accelerated erosion can be seen as an adjustment of sediment entrainment and transport mechanisms to altered ecological and pedological conditions. Soil structure is emphasised, as it is a key control on the pathways followed by water and sediments.

Geomorphology and ecology

Interactions between geomorphology and ecology can be approached at various spatial or temporal scales. These are most readily described concisely for temporal scales too short to register climatic changes and for spatial scales characteristic of specific vegetation associations. The chapter focuses on the effects of these interactions on soil structure and on the subsequent movement of water and soil particles through and over the soil. Attention centres on the way in which the composition of natural or human-controlled ecosystems influence the processes of water movement and sediment displacement, rather than on the more general effects of 'vegetation cover'.

Three themes, or precepts, form the background to linkages between geomorphological processes, soil structure, and organisms: (i) It is the structure of the soil which determines the pathways followed by run-off and the potential for erosion. In particular, the size distributions of pores and (micro-)aggregates, and the stability of the structural units, influence water acceptance and hydraulic conductivity. (ii) Processes operating in the soil at a microscopic scale can, if changed, alter the hydrological response of an area of 1 m^2, a hill slope, or a drainage basin. (iii) By redistributing rainfall, the vegetation cover, and in some cases other organisms too, can induce extremes in rainfall intensities and soil moisture conditions in excess of those suggested by meteorological data collected under standard conditions. These extremes can dominate the hydrogeomorphological response of an area.

The first two precepts are exemplified by the phenomena of soil sealing or crusting. Figure 5.1 illustrates infiltration measurements from a field area in the semi-arid Rif Mountains of Northern Morocco. Surface sealing increases the frequency of ponding and run-off, and results in an areal redistribution of the precipitation. This redistribution is of great significance for land management as well as for erosion. The occurrence of a surface crust at the sites referred to in Figure 5.1 is partly related to land management but also to the presence of small amounts of pedogenic oxides and organic material which affect the aggregation of the soil. The third precept, although particularly relevant to the occurrence of overland flow beneath the drip points of forest trees, is evident in the infiltration measurements from Morocco (Figure 5.1). Infiltration beneath a remnant of semi-natural pine forest some kilometres away from the other sites illustrates the effect of water repellency on the redistribution of rainfall. Very high infiltration rates for water entering the upper few centimetres of the litter contrast with the complete lack of infiltration into the water-repellent material on which the litter rests. Water is removed very rapidly by through-flow in the upper layers of the water-repellent soil. This phenomenon has also been observed in Southern Spain. It may increase the erosion occurring downslope from the forested areas where the water from the water-repellent soils emerges as return flow. This water repellency is, of course, temporary and decreases

Figure 5.1: Measurements of accumulated infiltration for adjacent highly calcareous non-crusted (1A) and crusted (1B) soils near Koubi, Morocco. The soil under pine forest was water repellent at the time of measurement.

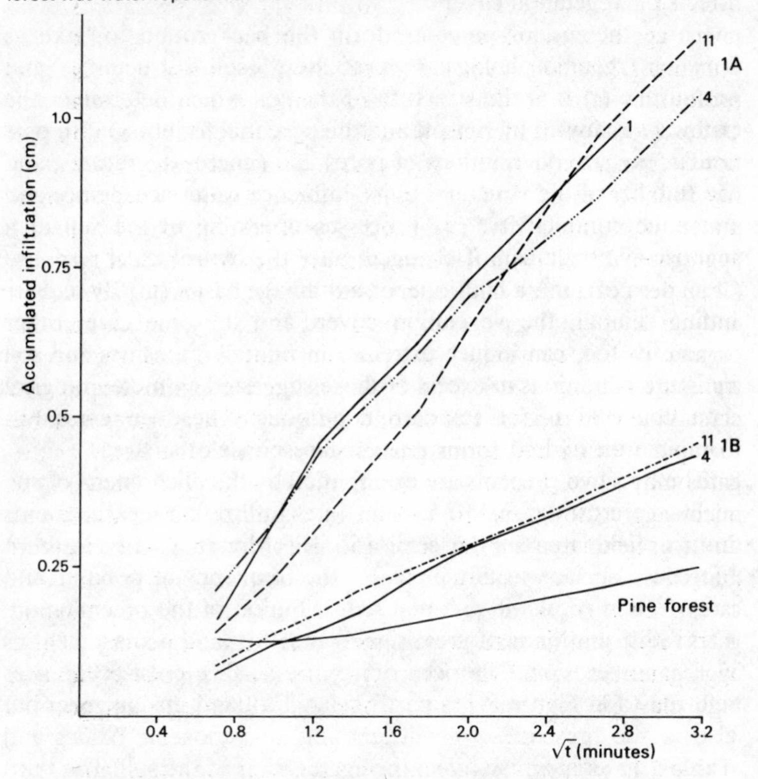

as the contact time with moisture increases, but it does significantly redistribute rainfall.

Soil aggregation, infiltration and erodibility

An understanding of the importance of soil structure in determining the pathways of water movement over and through the soil depends on some familiarity with aspects of soil aggregation related to infiltration and erodibility.

Soil aggregation

The structural elements of soil aggregates and the pore size classification suggested by Greenland (1979) are shown in Table 5.1. In general, soil aggregates are considered to be formed by agglomerations of microaggregates (< 250 µm) and skeleton grains. The micro-aggregates are composed of primary particles and clay domains (Emerson, 1959). The stabilisation of soil aggregates is profoundly influenced by the effect of organic matter. Tisdall and Oades (1982) consider the size distribution of stable agglomerations of particles in a red-brown earth from Australia. They identify four stages of aggregation, applicable to soils where organic matter is the main binding agent. The largest structural units are aggregates > 2000 µm. Where organic carbon contents are high (> 2 per cent) these large aggregates consist of smaller aggregates and particles held together by roots and fungal hyphae. Where the organic carbon content is < 1 per cent they are held together by transient binding agents and by disordered aluminosilicates and crystalline iron oxides. The second category of aggregates is stable to rapid wetting and forms particles into which the larger aggregates may slake. In the soils reported by Tisdall and Oades these micro-aggregates were 10-150 µm in diameter. Their stability is due to several types of strong binding agents (persistent organic materials, crystalline oxides and highly disordered aluminosilicates). These micro-aggregates are, in turn, composed of aggregates 2-20 µm in size which form the third category. These aggregates are bound very strongly by persistent organic bonds and high clay and organic matter contents. The fourth category of

Table 5.1: Terminology of soil pores according to Greenland (1979) and organisation of a soil aggregate from a red-brown earth (Tisdall and Oades, 1983)

Equivalent cylindrical diameter (µm)		diameter (µm)	
> 500	Fissures	> 2000	Water-stable aggregates bound by transient binding agents or fine roots and hyphae
50-500	Transmission pores	20-250	Micro aggregates stable to rapid wetting
0.07-50	Storage pores	2-20	Organic-rich water-stable aggregate
0.005-0.07	Residual pores	0.2-2	Floccules of clay plates
< 0.005	Bonding pores	< 0.2	Clay plates

aggregates, < 2 μm in diameter, consists of fine material held together by organic matter, iron oxides, and floccules of clay plates.

The dynamic nature of water-stable soil aggregation is important. The proportion of water-stable aggregates in a soil is continually changing since organic materials binding soil particles are continually being synthesised and decomposed. An example of the effect that these changes have on soil erodibility is illustrated for cultivated and forested soils in Luxembourg (Figure 5.2). The dynamic nature of aggregation, however, makes a simple index of soil-aggregate stability difficult to derive.

The infiltration envelope

The infiltration of water into a soil is closely related to aggregation and pore size distributions. Two situations can be usefully distinguished. The first is one in which infiltration is initially the result of soil matric forces, and then later, when suctional gradients become less, of gravitational forces. The second is when infiltration is predominantly accounted for by water moving through macropores, channels or pipes formed by volumetric changes in the solid phase or by biologic activity.

A useful approach to the first situation is the concept of the *infiltration envelope* which describes the relationship between rainfall intensity and the time to, or amount of rain required for, ponding (Smith, 1972). Infiltration envelopes can be compared with detailed pluviograph records to estimate the frequency of ponding. They are easy to establish from field measurements either directly or indirectly, using measured values of parameters required for the modified forms of the Phillip or Green and Ampt equations.

The infiltration envelope at a site is not constant, but varies with soil moisture content and the state of soil aggregation; certain local features can pose procedural problems for measurement (Bryan *et al.*, 1984). Nevertheless, ponding-time measurements are useful, as they enable the water-acceptance rates of soils to be compared. The ponding times shown in Figure 5.3 indicate how soil crusts greatly lower rates of water acceptance. The crusts in the calcareous soils tested were produced by raindrop impact and by slaking, and have rates of water acceptance only one-third those of freshly cultivated soil.

The second situation is gradually receiving more attention. A high degree of biological activity in the soil or shrinking and swell-

Figure 5.2: Seasonal variations in the erodibility of soil indicated by laboratory rainfall simulation experiments with undisturbed samples. The samples are from Al or Ap horizons of soils in Keuper marls under forest (1 and 3), pasture (2) or cultivation (4) (Imeson and Vis, 1984).

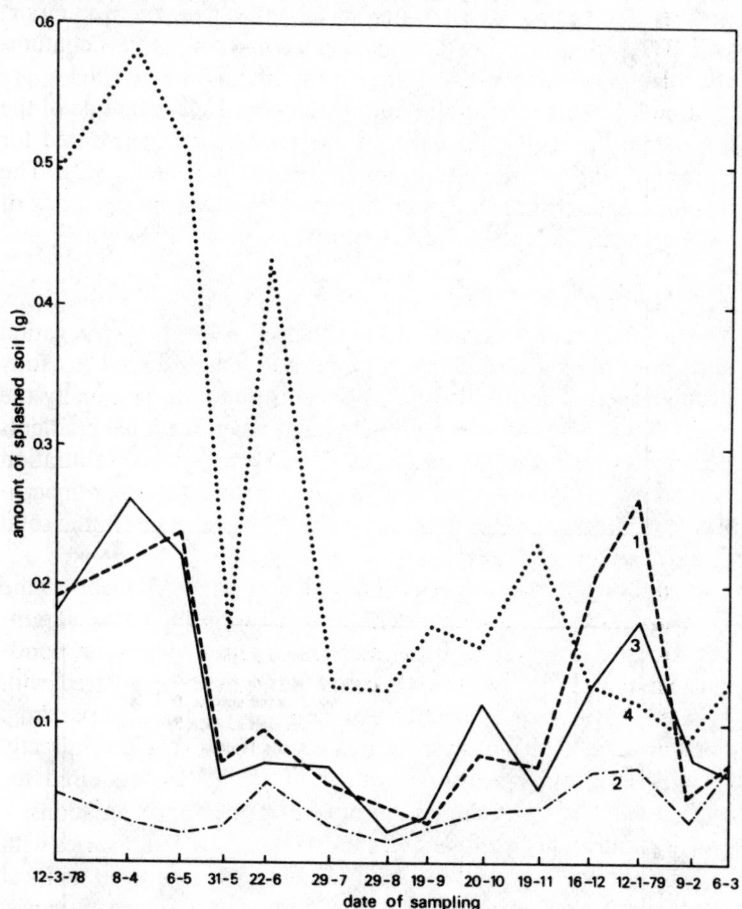

ing can greatly influence infiltration rates. Macropores are, however, relatively easily blocked by entrapped air. Therefore, the nature of the soil-air contact zone should be considered, since the roughness of the surface influences air entrapment ahead of the wetting front (Dixon and Simanton, 1979). Air venting at microtopographic highs is important in preventing entrapped air from building up relatively high pressure zones under microtopographic low points or furrows.

Figure 5.3: (a) Infiltration envelopes measured for different sites at Koubi, Morocco. (b) Infiltration envelopes for highly calcareous soils on an alluvial fan and on weathered marls at sites in southeast Spain. The vertical dashed line indicates the maximum ponding times for crusted soils (after Imeson and Verstraten, 1985).

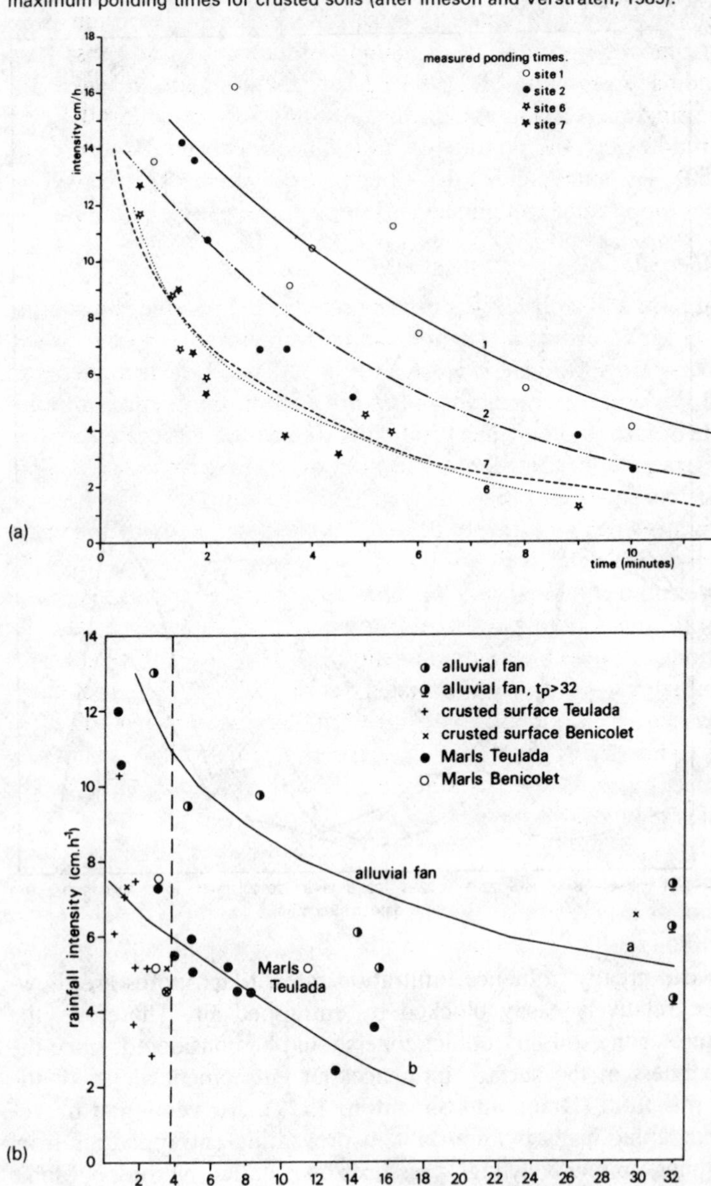

Roughness is also important since the threshold required for ponding is not the same as the threshold required for overland flow (Moore *et al.*, 1980). In Morocco, for example, it was found that the roughness characteristics of experimental erosion plots were more important in explaining differences in soil loss than ponding times (Imeson, 1983). Many methods are available for assessing microtopographic roughness and for predicting the time to runoff once the ponding point has been reached (Moore *et al.*, 1980). The simple index described by Seginer (1971), which combines topographic roughness and slope angle, is easy to apply.

Soil erodibility

The term *soil erodibility* is used to refer to the resistance a soil has to erosion, expressed by the actual amount of erosion which occurs. However, there are relatively few good measurements of soil loss and, when soil loss rates are known, they reflect a multitude of factors other than erodibility. Indices of aggregate stability, for example, cannot be used to indicate a high erodibility without first considering other factors. Soïl conditions which enable saturated overland flow to develop can lead to the erosion of easily transportable but highly water-stable aggregates. This is the case, for example, with certain volcanic ash soils described for Colombia. Removing the forest cover or creating clearings at certain elevations can lead to the permanently moist soils, which contain the unusual clay mineral allophane, irreversibly drying into easily splashed or entrained aggregates which have a very low bulk density. This drying results in an increase in infiltration and has an impact on slope stability since the water-retention capacity of the soil is reduced.

Soil erodibility involves two components: (i) the water-acceptance characteristics of the soil which are related to the generation of overland flow or through-flow; and (ii) the resistance of a soil to detachment and transport. Aggregate stability is related to both in a variety of ways depending on soil type, vegetation and topographic position. Useful possibilities for characterising aggregate stability exist (Bryan, 1968). One characterises the way in which soil particles *slake* (or disperse) when placed in water with or without a simple pre-treatment (Emerson, 1967; Loveday and Pyle, 1973). Another characterises erodibility by the amount of dispersible clay or silt present in a soil. This can be determined by slightly modifying the procedures used for grain-size analysis, but performed without pre-treatment to remove organic matter or carbonates (Chittleborough, 1982), or by measuring the turbidity of suspensions (Reid *et al.*, 1982). The water-drop test simply involves recording the response of individual aggregates to the impact of water drops having a known kinetic energy (Low, 1965). Perhaps the best-known index of erodibility is the k factor of the Universal Soil Loss Equation (Wischmeier and Smith, 1978), which refers to the erodibility of the soil influencing soil loss over a long period of time. As such it is difficult to compare with indices of erodibility or aggregate stability obtained from studying soil properties.

Plant roots

Plant and tree roots strengthen and stabilise soil material and slopes, and Gray and Leiser (1982) have indicated how this information can be applied in soil conservation in many ways. Roots of different species excrete substances which either positively or negatively influence aggregate stabilisation (Reid and Goss, 1981). Maize roots, for example, can destabilise a soil. More commonly, root stabilisation of large soil aggregates adds to that provided by the yearly addition of organic matter to soils by root growth and decay.

A case study from Luxembourg

Views of phenomena as intricate and locally variable as those controlling sediment movement remain out of focus unless scrutinised in a specific, real-world situation. The following case study, therefore, elucidates the linkage between soil structure and processes in 11 small watersheds in Central and Southern Luxembourg. Here, Keuper marls form a gently rolling landscape with a relief of about 60 m. The marls have generally weathered into heavy clay soils which have a silty surface horizon about 10-25 cm thick. About half of the region is forested with mixed oak (*Quercus cf robur, L.*) beech (*Fagus sylvatica, L.*) and hornbeam (*Carpinus betulus, L.*). Most of the remainder is under pasture or cultivation. The annual precipitation is about 780 mm, evenly distributed throughout the year.

Sediment and water budgets of forested and non-forested basins

From investigations with my colleague M. Vis, in 11 small watersheds it emerged that although forested and cultivated drainage basins produced similar annual amounts of run-off, flood peaks expressed in terms of discharge/km^2 (specific discharge) in the rivers under forest were twice as high as those from the cultivated catchments. The sediment yields of the cultivated drainage basins were nevertheless higher, being on average five times greater than in the forested catchments. Solute loads from the cultivated areas were, in contrast, only 30-50 per cent higher than for the forested areas.

The importance of macroporosity

In the forests, the higher storm run off (Figure 5.4) is generated from microtopographic depressions during the winter and spring, when a perched water-table is present in the B horizon of the soil. Due to the low permeability of the B and C horizons, this perched water-table extends over entire drainage basins at a depth of about 25-40 cm (Bonell *et al.*, 1984). Above the B horizon, the A horizon can be subdivided into a upper A_{11} horizon about 15 cm thick which has a very high biological activity, organic matter content and macroporosity, and a transitional A_{12} horizon. During wet weather, the water-table may rise locally to the surface for very short periods to produce saturated overland flow or through-flow in the A_{11} horizon. The through-flow in the A_{11} horizon is very rapid, following macropores or shrinkage cracks which remain open until December or January. The macropores form a network of drains, and continually change position due to the reworking of the soil by the macro- and meso-soil fauna. In detail, the perched water-table rises to maximum elevations which range from 0 to about 25 cm from the surface because the macroporosity and consequently drainage rates are so variable. Consequently, water may emerge on the surface for short distances as saturated overland flow at places where biopores are smaller or less frequent, only to infiltrate later where macroporosity increases. The pattern of macroporosity is controlled by the dynamics of the earthworm population and by predatory rodents. Earthworm activity in these forests is influenced by the composition of the forest.

The rapid biopore drainage cannot occur where cultivation disrupts the continuity of the macropores, nor under conifer planta-

Figure 5.4: Flood hydrographs from forested Robisbaach and Keiwelsbaach, and the cultivated Mosergriecht, which drain the Steinmergelkeuper formation of central Luxembourg.

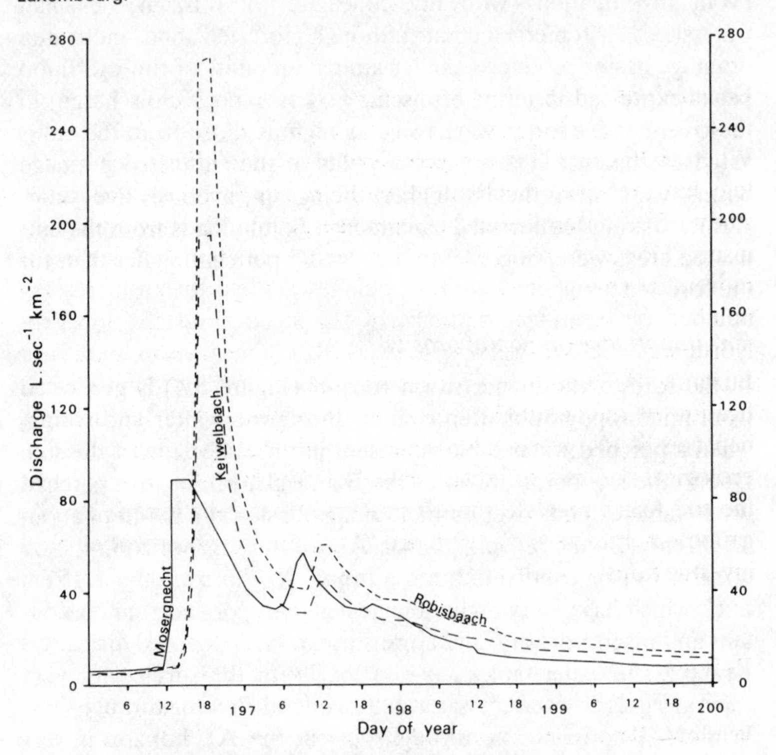

tions where acidity and the needle litter inhibit earthworms.

The linkage between soil structure and run-off

First, the higher peak discharges in streams draining forested basins means that the river channels are incised and are actively eroded. These channels show many characteristics associated with recent downcutting. In cultivated basins, river channels are now relatively much smaller and are often being aggraded. Secondly, the water which moves as through-flow takes up physically dispersed clay from the upper B horizon. Under cultivation this downslope translocation of clay through macropores cannot occur so that the sediment budget of the drainage basins is altered, as are the processes of soil development. Sediment concentrations nevertheless remain higher in run-off from the cultivated basins due to the much lower erodibility of the forest soils (Imeson and Vis, in press).

Conclusions

Whereas linkages between geomorphology and climate or hydrology have received much attention, there has been little systematic study of the interrelationshps between geomorphology and organisms. However, relationships between vegetation and sediment movement (budgeting) are now being described for an increasing number of locations, particularly for steep forested slopes in North-west USA (Swanson *et al.*, 1982). The present studies in Luxembourg continue to progress in their recognition and quantification of sediment transportation processes and of the linkages between soil structure and other storage elements. It seems that ecological controls so greatly influence the hydraulic properties of the soil which regulate run-off processes that their further investigation emerges as a major theme of continuing geomorphological investigations.

References

Bonell, M., Hendriks, M.R., Imeson, A.C. and Hazelhoff, L. (1984), 'The generation of storm runoff in a forested clayey drainage basin in Luxembourg', *J. Hydrol.*, *71*, 53-77.

Bryan, R.B. (1968), 'The development, use and efficiency of indices of soil erodibility', *Geoderma*, *2*, 5-26.

Bryan, R.B., Imeson, A.C. and Campbell, I. (1984) 'Solute-release and sediment entrainment on microcatchments in the Dinosaur Park badlands, Alberta, Canada', *J. Hydrol.*, *71*, 79-106.

Chittleborough, D.J. (1982), 'Effect of the method of dispersion on the yield of clay and fine clay', *Aust. J. Soil. Res.*, *20*, 339-46.

Dixon, R.M. and Simanton, J.R. (1979), 'Water infiltration processes and air–earth interface conditions' in *Proc. 3rd Int. Hydrol. Symp.* (Surface and Subsurface Hydrology), 27-29 June 1977 (Fort Collins), pp. 314-66.

Emerson, W.W. (1959), 'The structure of soil crumbs', *J. Soil Sci.*, *10*, 235-44.

Emerson, W.W. (1967), 'A classification of soil aggregates based on their coherence in water', *Aust. J. Soil Res.*, *5*, 47-57.

Gray, D.H. and Leiser, A.T. (1982), *Biotechnical Slope Protection and Erosion Control* (Van Nostrand Reinhold, New York).

Greenland, D.J. (1979), 'Structural organization of soils and crop production', in Lal, R. and Greenland D.J. (eds.), *Soil Physical Properties and Crop Production in the Tropics* (John Wiley, New York), pp. 47-56.

Imeson, A.C. (1983), 'Studies of erosion thresholds on semi-arid areas: Field measurements of soil loss and infiltration in Northern Morocco', *Catena Supp.*, *4*, 79-89.

Imeson, A.C. and Vis, M. (1984), 'Seasonal variations in soil erodibility under different land use types in Luxembourg', *J. Soil Sci.*, *35*, 323-31.

Loveday, J. and Pyle, J. (1973), 'The Emerson dispersion test and its relationship to hydraulic conductivity', *CSIRO Div. Soils Tech. Pap.*, *15*, 1-7.

Low, A.J. (1965), 'The study of soil structure in the field and laboratory', *J. Soil Sci.*, *5*, 57-74.

Moore, I.D., Larson, C.L. and Slack, D.C. (1980), 'Predicting infiltration and micro-relief surface storage for cultivated soils', *Univ. Minnesota Water Resour. Res. Cent. Bull.*, *102*.

Reid, T.B. and Goss, M.J. (1981), 'Effect of living roots of different plant species on the aggregate stability of two arable soils', *J. Soil Sci.*, *32*, 521-41.

Reid, T.B., Goss, M.J. and Robertson, P.D. (1982), 'Relationship between the decreases in soil stability effected by the growth of maize roots and changes in organically bound iron and aluminium', *J. Soil Sci.*, *33*, 397-410.

Seginer, I. (1971), 'A model for surface drainage of cultivated fields', *J. Hydrol.*, *13*, 139-51.

Smith, R.E. (1972), 'The infiltration envelope: results from a theoretical infiltrometer', *J. Hydrol.*, *17*, 1-21.

Swanson, F.J., Janda, R.J., Dunne, T. and Swanston, D.N. (1982), 'Sediment budgets and routing in forested drainage basins', *US Dept Agr. Forest Serv., Gen. Tech. Rep.*, PNW–141.

Tisdall, J.M. and Oades, J.M. (1982), 'Organic matter and water-stable aggregates in soils', *J. Soil Sci.*, *33*, 141-63.

Wischmeier, W.H. and Smith, D.D. (1978), 'Predicting rainfall erosion losses — a guide to conservation planning', *US Dept Agric. Hand.*, *537*.

6 THEMES IN AUSTRALIAN FLUVIAL GEOMORPHOLOGY

R. Warner

Introduction

Rivers are nature's gutters which transport water and sediment from high points to low in hierarchial fluvial systems. The water, largely derived from atmospheric sources, is dependent on climate, while the sediments are the products of weathering often reworked by running water. Where precipitation exceeds evapotranspiration, some water and part of the sediment load may reach adjacent oceans.

However, in Australia, which is large and low-lying, rain falling on about two-thirds of the land never reaches the sea, and sediment conveyed by temporary streams is stored in interior basins. This is a land of many rivers, most of which are dry for most of the time. Boat races on the Todd River, Alice Springs, are held on foot and it is quite startling when there is water in the river. Rivers in the arid and semi-arid centre are dry for most of the time; those flowing from wetter eastern divides into arid areas often dry up; seasonal droughts in the South (summer) and in the North (winter) can cause all but the largest streams to stop flowing; and long-term drought can affect even the moister coastal areas. Consequently, river regimes in Australia are notoriously erratic.

The scientific study of rivers is still somewhat limited, there being many rivers and few geomorphologists. However, hydrologists and engineers with state and federal authorities have gauged rivers extensively for water resource assessment. Practicalities precede scientific curiosity and many geomorphic studies remain applied in context. To the layman, perhaps, most rivers look alike. In fact, all are unique in that their channels reflect the integrated impacts of water and sediment discharges operating in different catchment sizes, and characterised by unique combinations of geology, soils, relief, vegetation, land use and occupance patterns. Even landform histories have often been quite different.

In this chapter, the main objective is to explore certain themes which illustrate the distinctiveness of Australian rivers. Such themes include the origins and their more recent natural evolution, their geography, the nature of changes induced both naturally and by man in 200 years of European settlement, and denudation rates. Discussion of such topics shows how Australian rivers are different, both within and outside this southern continent.

The origins

Australia is an ancient, stable landmass. The low-lying plateaux of the Western Shield feature mainly Precambrian rocks of great stability. In the East, the uplands are composed of much altered and planed Lower Palaeozoic lithologies. Between these two older masses, and peripheral to them in places, are areas of more recent deposition. Late Mesozoic and Tertiary earth movements have often enhanced local relief.

With the break-up of Gondwanaland, Australia began its migration from polar towards equatorial latitudes (Pillans, 1983). However, low elevations and general stability allowed little real change. Ancient drainage from the West was then mainly radial, but it lost continuity and coordination as the area became drier in what are now subtropical, high-pressure latitudes. East-flowing drainage from the Shield joined western contributions from the Highlands, allowing Central Mesozoic sedimentation. This has now been raised to provide the continuous land mass of the Central Lowlands. Drainage integration from the eastern uplands has been retained because of greater run-off from higher elevations (e.g. into the Lake Eyre Basin and in the Murray–Darling).

Individual drainage lines and their divides have been influenced by differential uplift which has created highs and lows on previously deeply-weathered surfaces. These surfaces and their evolution have received much attention (for reviews, see books edited by Jennings and Mabbutt, 1967; Davies and Williams, 1978; Langford-Smith, 1978). The apparent unity of the Eastern Highlands caused Andrews (1910) to suggest that there had once been a common extensive surface which was uplifted at the end of the Tertiary in what he called the Kosciusko Uplift. This notion, and those derived about drainage evolution from the examination of topographic maps, ran counter to geological evidence already available (Bishop, 1982). The refinement of early ideas about Tertiary rivers in Eastern Australia has come about with the dating of widespread, fossilising basalt flows, the determination of fossil flow directions, and the use of other geological techniques (Ollier, 1978; Young, 1981; Bishop, 1982). The general consensus now emerging is that divides have moved very little, making the origins of eastern and western drainage very old, and uplifts have either occurred early in the Tertiary or not at all. In the far West, there are fewer studies but it appears that ancient drainage lines have been disrupted by wind-borne sand movements, as climates became drier and run-off less effective.

More recent drainage evolution

In the Tertiary there were long periods of relatively stable climates with perhaps much more constant regimes. However, throughout the Pleistocene and Holocene there have been alternating climates, cooling in glacials and warming in interglacials. These, plus associated shifts in sea level, have been responsible for regime and base-level changes. Falls in sea level in glacials rejuvenated lower parts of coastal streams; rises in interglacial transgressions drowned lower parts.

Low latitudes and generally low altitudes meant that only small parts of Australia (in Tasmania and the South-east Highlands) were glaciated. Other parts of the uplands were subject to periglacial activity. Such processes, like those associated with glacial outwash, added to nearby fluvial loads, thereby modifying the character of some flood plains (or alluvial sinks) (Galloway, 1965).

Frequently, changing regimes meant that flood plains were abandoned to become terraces. Three kinds of climatic terraces have been described for coastal valleys (Figure 6.1): incised, where cut down into bedrock has been progressive; inset, where similar terraces have been formed in the same bedrock trough; and overlapped, where transgressions or heavier sedimentation have buried older flood plains under recent sediments (Warner, 1972). Such studies have been limited to Victoria, NSW, and Western Australia (Walker, 1969; Hickin, 1970; Warner, 1972). Dateable carbonaceous materials have restricted an absolute chronology to about 30 000 years. Also, in narrow coastal bedrock valleys, the survival of older flood plains has been limited. On the Riverine

Figure 6.1: Three types of climatic terraces in coastal valleys.

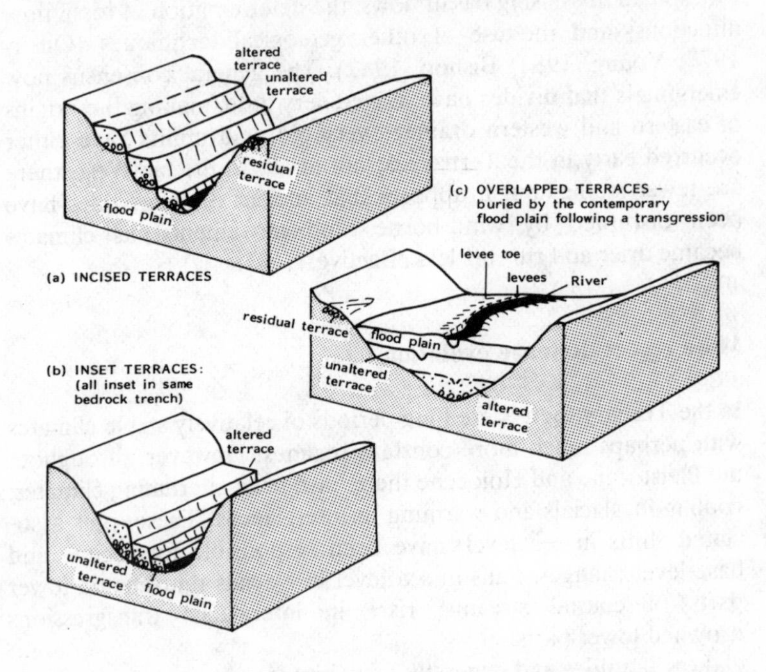

Plains, much more stratigraphic evidence is available (Bowler and Harford, 1966; Schumm, 1968; Butler *et al.*, 1973). Alternations of arid and fluvial processes have created complex landform assemblages including lakes, dunes, salinas, lunettes, ancient channels, anabranches, and so on. The exotic rivers crossing these dry plains have responded to climatic changes in their Highland catchments, as well as to those on the plains.

Three main stages of fluvial surface forms have been identified: prior streams, of 20 000 years of more, related to the last glacial stage; ancestral rivers of the Holocene; and the present channels (Schumm, 1968). Not all changes are related to climatic variations. Some adjustments have resulted from single events, like blockages caused by vegetation rafts, or from the effects of local faulting.

The geography of Australian rivers

Australia is a dry continent, with mean annual rainfall of only

420 mm and run-off depth of only 45 mm (Warner, 1977). Spatial and temporal variations, however, are high, with a wet periphery, notably from the North clockwise to the South-east, and a dry 'dead heart' which was pulsating strongly in the 'big wet' of January 1984.

The average condition is given by the run-off map (Figure 6.2), but the variability of flow in Australian rivers can never be over-exaggerated. Run-off, and therefore stream regime, is closely related to the amount, seasonality, duration and intensity of precipitation. Throughout much of the Centre the median rainfall is less than 250 mm and most of the continent (62.5 per cent) gets less than 400 mm. In contrast, parts of Western Tasmania and North-east Queensland have a median fall of over 2400 mm. Cold front rains dominate in the South in winter. In summer, these are pushed further to the South by southward-migrating high-pressure cells. This gives the dry Mediterranean-type summers for Adelaide and Perth. The North is dry in winter, being dominated by the dry, stable air of these same high-pressure cells. As these move southwards in summer, they allow incursions of moister equatorial air, which bring monsoonal, convectional and cyclonic downpours, mainly to the coastal zone. Orographic effects can be pronounced in the Eastern Highlands; elsewhere, local effects can be important, with the Perth Plain, west of the Darling Range, getting more than 1200 mm annual rainfall. To the east of this low barrier, rainfall soon drops below 400 mm. Thus, seasonality in rainfall is induced by high-pressure cell migrations: wet winters in the South, wet summers in the North, rain at any time in the East, and the dry Centre, with incursions from various sources to yield episodic falls.

Drainage types and distribution (Figure 6.2) can be based on divides between external and internal drainage, with the latter either coordinated or uncoordinated. Twelve divisions were proposed mainly for water resource assessment (Table 6.1). Eight are coordinated external divisions with direct run-off to the sea, varying in width from < 100 to > 500 km. The only coordinated internal division with an external outlet is the Murray–Darling, Australia's longest and most utilised river system. Its mean run-off is less than half the average and, in dry years, contact with the sea can be lost. The Lake Eyre division, which is tectonically depressed below sea level in the Centre, is an internal coordinated system where run-off from heavy rains in the Queensland uplands can flood this large salt lake. Water is removed by evaporation in

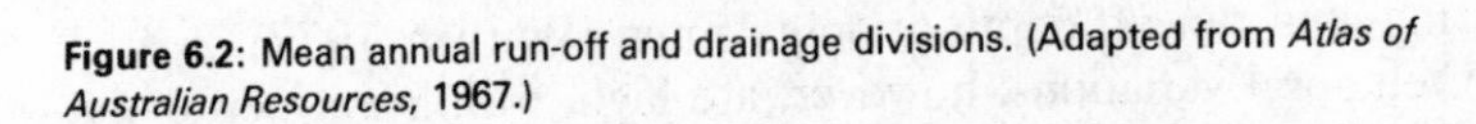

Figure 6.2: Mean annual run-off and drainage divisions. (Adapted from *Atlas of Australian Resources*, 1967.)

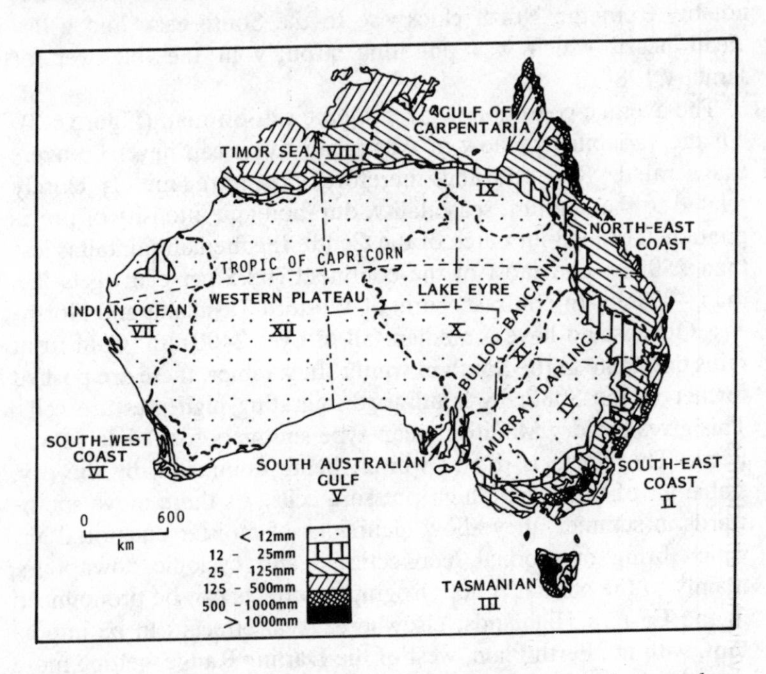

succeeding dry years. The rest is uncoordinated drainage where temporary drainage nets can be re-established in parts of earlier systems following heavy rains.

It is thought that certain magnitudes of discharge help influence how much sediment may be delivered to the channel. The geology of the catchment, as well as the closeness of sediment sources, will influence the size of material forming the channel's perimeter and therefore the shape of the channel (Schumm, 1977). It seems probable that denudation may be highest where run-off occurs in sparsely vegetated, semi-arid areas, and where temporary failure of vegetation (i.e. after the seasonal drought or in developed areas in the tropical savanna) causes rapid loss of sediments in subsequent high-intensity rainfall events (Williams, 1976; Pickup *et al.*, 1983).

For these reasons, there are many types of channel. Some rivers flow through several climatic zones and may therefore display several types along a single waterway. So, in a sense, there are many 'distinctive' regimes, most of which fail to show much regularity. For instance, the savanna rivers have highly seasonal

Table 6.1: Drainage divisions, areas, and run-off volumes and depths

			Area (km^2)	Mean RO (mill m^3)[a]	Mean RO depth (mm)[b]
A. External coordinated	VII	Timor Sea (summer max. rain)	539 000	74 287	138
	IX	Gulf of Carpentaria	640 800	63 146	99
	I	North-east coast	454 000	82 867	183
	II	South coast (mixed seasonal rain)	268 000	36 400	136
	III	Tasmania	68 400	47 171	690
	V	S. Australian Gulf (winter max. rain)	75 400	532	7
	VI	South-west coast	140 000	7 223	51
	VII	Indian Ocean (arid coast rain)	520 000	6 160	12
B. Internal/external	IV	Murray-Darling (summer/winter low)	1 057 000	23 663	22
C. Internal coordinated	X	Lake Eyre (semi-arid, arid)	1 144 000	3 906	3
D. Internal disconnected	XI	Bullo-Bancannia (arid)	100 800	407	4
E. Uncoordinated	XII	Western plateau (arid)	2 679 000	—	—
		Totals	7 686 370	345 762	45

Source: AWRC (1965).

[a] Mean run-off (total in millions m^3).

[b] Mean run-off depth over whole catchment areas.

(summer), intense rainfall, and denudation is highest at the beginning of the wet season, when the surface is least protected. Tropical rivers exist more in the North-east, where rain and run-off occur nearly all year, with maxima in summer. Dense vegetation restricts the denudation rate. Sub-tropical rivers are similar, but the dominance of vegetation, rainfall and run-off are all less. Temperate rivers in the East may flow all year, and vegetation, except where disturbed on a large scale, can maintain denudation at low levels. The Mediterranean rivers of the South have low run-off, mainly in winter. In the driest areas, rainfall and run-off events are very much more episodic and the relative absence of vegetation may promote moderate (albeit infrequent) denudation.

It is also possible to use other criteria to define different channel types. For instance, Schumm (1977) has suspended-, mixed- and bedload streams; while Pickup (1984) has defined source reaches, where the channel is in contact with bedrock; armoured reaches, where the bed is protected by a veneer of coarse bedload material; mobile zones, where bed material is fine enough to be transported by large flows; and sink zones, where backwater effects from larger streams or coastal systems may cause deposition of fines on the bed. Consequently, short coastal streams are mainly mixed- or bedload channels, especially away from sandstone lithologies. Much of their length may be in incised bedrock meanders of the source zone, with tongues of armoured reaches extending often to near deltaic distributaries, which are suspended-load channels and sediment sinks. Such streams display a normal hydraulic geometry, with tributaries adding more water and sediments in a downstream direction. Floods in such channels can occur at any time in the East, but are largely confined to summer in the North and winter in the South (or spring with snowmelt from the coldest areas).

Internal coordinated drainage derived from the Eastern Highlands commences as mixed- and bedload drainage in source reaches. Where coarse loads are added, armoured channels extend onto the plains. These then give way to long, mobile sandbed reaches, and eventually to very fine suspended-load channels in lower mobile and sink reaches. Gradients are very low, giving the reticulate forms in the Queensland Channel Country and anabranching in the Murray–Darling (Dury, 1969). Hydraulic geometries are not normal because discharge decreases in a downstream direction (Schumm, 1968). Thus, while slope and roughness may decrease (which is 'normal'), width, depth (perhaps), velocity and sediment load may also decrease. Such rivers flow long distances in dry country and water may be 'lost' to influent drainage and evaporation.

Much less work has been attempted on the uncoordinated systems of the West. The presence of source and armoured reaches is common, together with rapid fining as temporarily mobile loads are deposited in flood-outs and interior basins. Chains of salinas often characterise ancient drainage systems in Western Australia which act as barriers against coordinated drainage.

Natural regimes and their changes

Hydrological regimes describe the amount and timing of water passing through river systems. A regular regime would exist where the timing and size of events were similar on the annual hydrograph (e.g. snowmelt regime), whereas a stable regime would show no progressive deviations over a period of years. Such regularity or stability never quite exists, as magnitude frequency studies reveal — large events with low frequency and smaller ones being more common occurrences. However, where variability is low and seasonality is pronounced, a stable regime is assumed to produce a stable channel morphology. In Australia, it has often been assumed that channels are or have been stable and are related to the operative regime, but demonstration of this is not easy because most channels have erratic regimes and display considerable instability.

Awareness of such problems has come through the study of regime changes through time and channel responses (Pickup, 1975; Erskine and Bell, 1982). Attempts to show relations between flow phenomena and morphology have been attempted (e.g. Woodyer, 1968; Pickup and Warner, 1976). Morphological considerations were based on benchfull, bankfull, statistically derived most probable annual floods and computed most effective discharges in channels, both stable and unstable, and over time-periods where there had been a marked change in regime (Pickup, 1976). This period (1945-78) is now thought to represent a secular change in climate in central East–West areas. The magnitude and incidence of flooding increased markedly, with rainfall increases of up to 30 per cent and more on 1900–45 means. Impacts of this on 'channel-forming' discharges (if assumed to be $Q_{1.58}$ or most probable annual flood (Dury, 1969) were marked by a sixfold increases at Wallacia on the Nepean River (Pickup, 1976). More general impacts of this and other changes have been reviewed (Erskine and Bell, 1982; Warner, 1984).

In the last century, between 1867 and 1900, there was another period of high flood activity in eastern New South Wales, but few flow records exist (Riley, 1981). Rainfall figures and some stage-and-flow observations are available for the Nepean River (Warner, 1983). The greatest event was in 1867, followed by another large flood in 1900. Both these events were considerably larger than any in the 1945–78 period, but channel adjustments in this period were never recorded. A 1900 survey for part of the Nepean above Penrith was taken after that flood. It undoubtedly included impacts of that event, because subsequent air photographs revealed a considerable reduction of width over the next four decades. Thus in the 200-year occupance period there have been considerable changes in regime. If channel-forming discharges exist between most effective and bankfull (Q_{me}-Q_{bf}), there must have been many changes in size and shape of channels. Changes for the last wet period are easier to document because old air photographs and surveys show pre-change channel morphologies. Very few of these systems are natural, however, and they incorporate the impacts not only of natural changes, but also those induced by man.

Channel changes: man's impact on regimes

Catchment changes can modify net run-off and also the sediment yield (Gregory and Walling, 1979). Most of such alterations have taken place in the last 200 years (Warner, 1984). Initially they involved forest clearance for firewood and building materials, as well as for grazing and cropping. In the early days this was a piecemeal operation around penal settlements and what are now state capitals. Affected areas increased as the occupance frontier expanded both into forested and arid regions. Grazing has probably had the greatest impact in the catchment area because it has been so widespread. This involved not only useful animals but also wild and discarded species. Rabbits, wild horses, camels and wild pigs have all posed major problems, and native grazing animals pose a threat to improved crops and fences. Over-grazing and drought put real stress on catchments and can lead to erosion in the catchment, as well as adjacent channel instability.

Systematic studies of catchments have not been numerous and those involving conservation have concentrated on slope responses to run-off and sediment yield under various forms of treatment (e.g. Adamson, 1974). Seldom do such investigations include responses in adjacent channels. Large-scale deforestation took place on the plains, on the table lands and in the coastal valleys long before any attempt was made to understand relevant impacts. Scientific studies have been much more recent and have been directed towards understanding the effects of clear-felling operations for woodchip industries (Western Australia — Conacher, 1977; New South Wales — Olive and Walker, 1982).

Deforestation did not always mean tree removal. Ringbarking was a way of killing trees and allowing grass to grow below them. Eventually, the trees would rot and could be burned. From such areas there was a higher run-off and sediment yield, especially with soil compaction. The impacts of this practice on tableland channels, known as 'chains of ponds', in the vicinity of Canberra were studied by Eyles (1977), who found that such channels which had been named and described by surveyors in the 1820s had now become deep gullies, the result of land clearance, soil compaction and grazing. The development of such gullies has been enhanced by urbanisation, and retarded by the application of fertilisers. The cropping of such land was difficult until tree stumps were removed and, if there were slopes, early ploughing was downhill rather than on the contour. Such actions have produced gullying in South Australia (Twidale and Bourne, 1978).

Other major catchment changes involve open-cast mining, urbanisation and conservation. In the case of mining, surface conditions may be greatly modified in areas mined and where wastes have to be stored. Environmental contamination can result from such piles, particularly if they contain radioactive wastes (Warner and Pickup, 1982; Pickup *et al.*, 1983).

Urbanisation is probably the most significant form of catchment alteration. More than 80 per cent of Australians live in urban areas, with the greatest effects in the large state capitals. The land surface is sealed by roads and buildings, and consequently infiltration is greatly reduced. Vegetation removal and loss of soil areas cut down evapotranspiration, while concrete, bitumen and lined stormwater drainage lines speed up the removal of increased run-off. Thus urban rivers become 'flashy' — that is, subject to frequent flooding — and have difficulty coping with increased run-off. These changes often require improved drainage which delivers water and sediments to nearby tidal estuary sinks much more rapidly (Warner *et al.*, 1977). The degree of urbanisation affects the volumes of run-off and sediment yields. Both are maintained at high levels in garden suburbs, whereas, although run-off is still high, sediment yield is much lower from more densely settled and commercial areas. Conservation measures, like contour banks, and water-harvesting projects, slow down water and sediment yields. The main idea is to retain water, nutrients and soils and to reduce thereby soil erosion, to increase infiltration and generally to protect the land surface.

Man's impact on channels are readily apparent at the affected sites, e.g. dams, dredging holes, weirs, channel improvements, bank stabilisation, and so on. However, such changes have repercussions up- and downstream, which are not always appreciated until problems arise. In recent years numerous studies have demonstrated the impacts of dams and weirs (Gregory, 1977; Warner, 1983, 1984); of dredging (Warner *et al.*, 1977); of channel-straightening for flood mitigation (Bird, 1980); and of urbanisation of channels (Nanson and Young, 1981). Changes subsequent to channel alteration can be assessed when there are earlier dimensional data for the channel.

Denudation

All rivers perform work in that they convey products of weathering and mass wasting from high points to low. Overland flows, outside the channel, operate in much the same way. Moving water carries dissolved loads of solution, finer sediments in suspension and coarser materials as bedload. Important in work rates are water volumes, their velocities and available materials, both from outside and in the channel.

Denudation rates can be worked out by knowing the volume of eroded material from the land mass in a given time; the volume of material deposited in a dam or delta, again in known time from a specific source area; by measuring the actual lowering of the landscape using erosion pins; and by computing the amount of material actually carried by the stream per year. This is done by working out sediment concentrations and volumes of water involved. These four approaches may be associated with geological long-term rates, intermediate to long-term rates, and contemporary rates. Denudation can be computed as mean catchment lowering in mm/1000 years, while present-day rates may be given in $m^3/km^2/yr$ or $t/km^2/yr$. The latter two can be converted to mm/1000 years if they are deemed to represent long-term rates. In areas now subject to accelerated rates through the intervention of man, such extrapolations are unwarranted.

Most denudation studies have been confined to the Eastern Highlands and to parts of the Murray–Darling catchment (Loughran, 1969; Douglas, 1973; Olive and Walker, 1982; Pillans, 1983). For the most part, these have involved small catchments and generally reveal low rates of denudation (10-40 mm/1000 yrs). These are low compared with mountainous regions and other continents (Olive and Walker, 1982; Pillans, 1983). In parts of North-east Queensland, Douglas (1973) found higher rates (mean $>$ 40 mm/1000 years) and these were related with heavy and high-intensity precipitation in forested areas. These were high when compared with about 8 mm for the southern tablelands, 9 mm for the drier northern tablelands of New South Wales, and over 30 mm for the very wet eastern margins of the latter. Compilations from radiometric ages and landscape reconstructions given in the same source vary from 5-30 mm/1000 yrs. Exceptions to these rates range from 1.2 mm for the 139 000 km^2 Barwon catchment, part of the Darling River, to 132-270 mm/1000 yrs for reservoirs in the Broken Hill area (rates given this way for comparison only) (Olive and Walker, 1982). Relatively high rates were also quoted for well-worked parts of the Hunter Valley (38-117 mm) and for a small granite, forested catchment in North-east Queensland (285 mm).

Detailed work in drier areas may well reveal fairly low rates of denudation, related to very low run-off and generally low slopes. However, land degradation, estimated at 32 per cent of all non-arid cropping and grazing land, has given cause for concern. The fact that such rates are not conveyed by the rivers may reflect generally very low sediment delivery ratios, with material being stored in various sediment sinks outside and in the channels. This view is supported by often very high sediment yields from small plots reflecting the impacts of land use, slopes, soil types and rainfall characteristics. Data given by Olive and Walker (1982, Table 3) reveal up to 15 280 $m^3/km^2/yr$ for a small hoed field at Goondi in

Queensland (more than 15 m/1000 yrs), from 2000-8000 m^3 for wheat/fallow rotations on sites west of the divide in New South Wales, 16 m^3 for improved pasture, and over 1400 $m^3/km^2/yr$ for unimproved pasture at Wagga Wagga.

Figures from the Adelaide River in the Northern Territory, based on a short record, reveal a mean denudation rate of only 7 mm/1000 yrs in the savanna, but it seems probable that the effects of the drought-breaking storms may not have been included (Williams, 1976). Such rates translated in sedimentation on the backwater Magela Plains reveal a mean fill rate of 500 mm/1000 yrs (Warner and Pickup, 1982). This has been partially confirmed by a 2000-year-old date for estuarine deposits at only 1400 mm below the plain surface (Pickup *et al.*, 1983).

It is evident that low rainfall in natural areas of low relief will produce a low denudation rate. This has characterised Australia both in the geological past as well as at the present time, as revealed by contemporary rivers. However, high rates of sediment removal are occurring outside of the channel and in gullies where poor land use and fire have been experienced.

Conclusions

Australia is a low-lying and stable platform, made up of western Precambrian plateaux, eastern subdued Lower Palaeozoic uplands and more recent sediments of the Central Lowlands. Lack of high mountains and sub-tropical, high-pressure latitudes help to make the continent very dry, particularly the Centre and West, where internal drainage predominates. Peripheral rivers which flow into the sea are generally short, drain catchments with relatively low, ill-defined divides, and have erratic regimes which depend to a large extent on the movement of high-pressure cells. Northern savanna rivers are dry all 'winter', but carry run-off in summer from the monsoon, thunderstorms and occasional cyclones. Southern mediterranean-climate rivers only flow in winter, while those of the East and South-east can flow in any season. Arid zone rivers are used only after episodic downpours. Thus rivers are distinctive, reflecting the impacts of different and often highly variable regimes. They have been inherited from relatively stable conditions of the Tertiary, through multiple changes of climate and external baselevels in the Pleistocene and Holocene. The types and positions of abandoned flood plains or terraces allow reconstruction of the last 30 000 years with some confidence. Change has continued in the Holocene and particularly in the 200 years of European occupance, both naturally with the impacts of short-term climatic changes and with man-induced modification in the catchments and channels.

Under natural conditions, denudation rates were and are amongst the lowest in the world. These have been derived from landscape reconstruction and from river loads, mainly in small catchments. However small plot observations in areas subject to land abuse, high intensity rains, duplex soils and steeper slopes indicate local erosion has been accelerated by at least two orders of magnitude. These and obvious land degradation suggest low sediment delivery rates to channels.

References

Adamson, C.M. (1974), 'Effects of soil conservation treatment on runoff and sediment loss from a catchment in South Western New South Wales, Australia', *Int. Ass. Sci. Hydrol. Publ., 113,* 3-14.

Andrews, E.C. (1910), 'Geographical unity of Eastern Australia in late and post-Tertiary time', *J. R. Soc., NSW, 67,* 251-350.

AWRC (Australian Water Resources Council) (1965), *Review of Australia's Water Resources: Streamflow and Underground Resources* (Dept. Nat. Dev., Canberra).

Bird, J.F. (1980), 'Geomorphological implications of flood control measures', *Aust. Geog. Stud., 18,* 169-83.

Bishop, P. (1982), 'Stability or change: a review of ideas on ancient drainage in Eastern New South Wales', *Aust. Geog., 15,* 219-30

Bowler, J.M. and Harford, L.B. (1966), 'Quaternary tectonics and the evolution of the Riverine Plain near Echuca, Victoria', *J. Geol. Soc. Aust., 13,* 339-54.

Butler, B.E., Blackburn, G., Bowler, J.M., Lawrence, C.R., Newell, J.W. and Pels, S. (1973), *A Geomorphic Map of the Riverine Plain of South-Eastern Australia* (ANU Press, Canberra).

Conacher, A. (1977), 'Conservation and geography: the case of the Manjimup woodchip industry', *Aust. Geog. Stud., 15,* 104-21.

Davies, J.L. and Williams, M.A.J. (eds.) (1978), *Landform Evolution in Australia* (ANU Press, Canberra).

Douglas, I. (1973), 'Rates of denudation in selected small catchments in Eastern Australia', *Univ. Hull Occ. Pap. Geog., 21.*

Dury, G.H. (1969), 'Relation of morphometry to runoff frequency', in Chorley, R.J. (ed.), *Water, Earth and Man* (Methuen, London), pp. 419-30.

Erskine, W. and Bell, F.C. (1982), 'Rainfall, floods and river channel changes in the Upper Hunter', *Aust. Geog. Stud., 20,* 183-96.

Eyles, R.J. (1977), 'Changes in drainage networks since 1820, Southern Tablelands, NSW', *Aust. Geog., 13,* 377-86.

Galloway, R.W. (1965), 'Late Quaternary climates in Australia', *J. Geol., 73,* 603-18.

Gregory, K.J. (1977), 'Channel and network metamorphosis in northern New South Wales', in Gregory, K.J. (ed.), *River Channel Changes* (John Wiley,. Chichester), pp. 389-410.

Gregory, K.J. and Walling, D.E. (eds) (1979), *Man and Environmental Processes* (Dawson Westview, Folkestone).

Hickin, E.J. (1970), 'The terraces of the Lower Colo and Hawkesbury drainage basins, New South Wales', *Aust. Geog.*, *11*, 278-87.

Jennings, J.N. and Mabbutt, J.A. (eds.) (1967), *Landform Studies from Australia and New Guinea* (ANU Press, Canberra).

Langford-Smith, T. (ed.) (1978), *Silcrete in Australia* (Dept. Geog. Univ. New England, Armidale).

Loughran, R.J. (1969), 'Fluvial erosion in five small catchments near Armidale, New South Wales', *Univ. New England, Res. Ser. Phys. Geog.*, *1*.

Nanson, G.C. and Young, R.W. (1981), 'Downstream reduction of rural channel size with contrasting urban effects in small coastal streams of Southeastern Australia', *J. Hydrol.*, *52*, 238-55.

Olive, L.J. and Walker, P.H. (1982), Processes in overland flow — erosion and production of suspended material', in O'Loughlin, E.M. and Cullen, P. (eds), *Prediction in Water Quality* (Aust. Acad. Sci., Canberra).

Ollier, C.D. (1978) 'Tectonics and geomorphology of the Eastern Highlands' in Davies, J.L. and Williams, M.A.J. (eds), *Landform Evolution in Australasia* (ANU Press, Canberra), 5-47.

Pickup, G. (1975) 'Downstream variations in morphology, flow conditions and sediment transport in an eroding channel', *Z. Geomorphol.*, *19*, 443-59.

Pickup, G. (1976), 'Geomorphic effects of changes in river runoff, Cumberland Basin, New South Wales', *Aust. Geog.*, *13*, 188-94.

Pickup, G. (1984), 'Geomorphology of tropical rivers: I. Landforms, hydrology and sedimentation in the Fly and Lower Purari, Papua New Guinea', *Catena*.

Pickup, G. and R.F. Warner (1976), 'Effects of hydrologic regime on magnitude and frequency of dominant discharge', *J. Hydrol.*, *24*, 51-75.

Pickup, G., Wasson, R.J., Warner, R.F. and Tongway, D. (1983), 'Geomorphic research for the long-term management of uranium mill tailings', CSIRO Rep., Alligator Rivers Region.

Pillans, B. (1983), 'Canberra, twenty million years on', *Aust. Geog. Stud.*, *21*, 92-101.

Riley, S. (1981), 'The relative influence of dams and secular climate change on downstream flooding, Australia', *Water Res. Bull.*, *17*, 361-6.

Schumm, S.A. (1968), 'River adjustment to altered hydrologic regimen — Murrumbidgee River and paleochannels, Australia', *US Geol. Surv. Prof. Pap.*, 598.

Schumm, S.A. (1977), *The Fluvial System* (John Wiley, New York).

Twidale, R. and Bourne, J.A. (1978), 'Distributions of relic "lands" or strip fields in South Australia', *Aust. Geog.*, *14*, 22-9.

Walker, P.H. (1969) 'Depositional and soil history along the lower Mackay River, NSW', *J. Geol. Soc. Aust.*, *16*, 683-96.

Warner, R.F. (1972), 'River terrace types in the coastal valleys of New South Wales', *Aust. Geog.*, *12*, 1-22.

Warner, R.F. (1977), 'Hydrology', in Jeans, D.N. (ed.), *Australia — a Geography* (Sydney Univ. Press, Sydney), pp. 53-84.

Warner, R.F. (1983) 'Channel changes in the sandstone and shale reaches of the Nepean River, NSW', in Young, R.W. and Nanson, G.C. (eds) *Aspects of Australian Sandstone Landscape* (ANZ Geomorph. Gp.) Spec. Publ., *1*, 106-19.

Warner, R.F. (1984, in press), 'Man's impacts on Australian drainage systems', *Aus. Geog.*

Warner, R.F., McLean, E.J. and Pickup, G. (1977), 'Changes in an urban water resource, an example from Sydney, Australia', *Earth Surf. Processes, 2*, 29-38.

Warner, R.F. and Pickup, G. (1982) 'The geomorphology of the wet and dry tropics and problems associated with the storage of uranium tailings in Northern Australia', *Uranium Mill Tailings Management, Proc. WEA Workshops,* (Colorado State Univ., Fort Collins, USA., pp. 45-58.

Williams, M.A.J. (1976), 'Erosion in the Alligator Rivers area', in Storey, R. *et al., Lands of the Alligator Rivers Area, Northern Territory, CSIRO Land Res. Ser., 38*, pp. 112-25.

Woodyer, K.D. (1968), 'Bankfull frequency in rivers', *J. Hydrol., 6*, 114-42.

Young, R.W. (1981), 'Denudation history of the South-Central uplands of New South Wales', *Aust. Geog., 15*, 77-88.

7 TROPICAL FLUVIAL GEOMORPHOLOGY

Richard H. Kesel

Introduction

The concept of morphoclimatic regions should be viewed cautiously by the prospective student of geomorphology. The concept is based more on a hypothetical than a realistic view of geomorphology. The morphoclimatic approach emphasises the distinctiveness of climatic regions. Workers within these regions have attempted to recognise geomorphic process and associated landforms that are unique to a particular environment (Tricart, 1973; Büdel, 1982). Tropical geomorphology represents one extreme within this conceptual framework. Arguments for the distinctiveness of landforms and processes within the tropical environment are based largely on the intensity of chemical weathering and its influence on denudational processes (Büdel, 1957; Tricart, 1973). Some landforms and processes indeed appear unique to the equatorial zone, such as beachrock formation and features related to coral- and mangrove-type coasts. Yet the distinctiveness of other landforms and processes, such as those associated with tropical karst, has been questioned seriously (Jennings, 1983). Furthermore, many soil scientists recognise no real difference in the weathering and soil-forming processes in the humid tropics as compared to humid temperate climates.

The prominent themes in tropical geomorphology include: (1) deep-weathering and duricrust formation; (2) planation surfaces including an emphasis on inselberg formation; (3) valley and slope formation; and (4) morphologic and climatic stability. These themes focus on denudational landforms rather than depositional ones. Knowledge of features like flood plains and alluvial fans has not kept pace with that from other climatic regions; therefore, the influence of the tropical environment on these landforms is uncertain.

This chapter reviews briefly some background ideas concerning tropical fluvial systems, including the effects of tectonic and climatic stability which may influence such systems. Subsequently, the discussion focuses on the development of alluvial fans in a tropical environment. Their comparison with fans in other environments is a way of assessing the distinctiveness of the tropical environment in fluvial depositional landform development.

Tropical fluvial systems

Tropical river systems are generally considered to be distinctly different from extra-tropical rivers (Tricart, 1973; Büdel, 1982). Tropical mountain valleys are described as narrow and V-shaped, lacking gravel in their channels. On tropical plains, rivers do not cut down to form valleys and generally exhibit little or no lateral erosion. Louis (1973) describes the longitudinal profile of tropical rivers as less concave with steeper gradients than similar rivers in mid-latitudes. Pippan's (1970) studies, however, based on measurements in Puerto Rico and the Bohemian Forest, concluded that valley profiles in the mid-latitudes and the Tropics have similar gradients. Schumm (1977, p. 25) indicates that downstream the depths of ephemeral channels increase and gradient decrease is less rapid than in humid streams, due to the high water loss in ephemeral channels. Büdel (1982, p. 150) and Tricart (1973, p. 58) note that tropical rivers have numerous rapids and waterfalls that alternate with reaches with low gradients, flowing over weathered rock.

Sediment transport and deposition by tropical rivers have received little attention, despite the number and variety of depositional environments found in the equatorial zone. Generally, rivers flowing in the humid tropical forest, if unaffected by human activity, have relatively low sediment concentrations. However, such areas may be susceptible to high-magnitude or catastrophic events, which may trigger episodes of high sediment yield. Rivers in the wet–dry savanna probably carry somewhat higher sediment loads. Because of the intensity and depth of chemical weathering, it is argued that tropical rivers have high suspended and dissolved loads, but little or no effective bedload and are therefore unable to erode their channels. Where rivers are found that contain gravel bedload, it is transported only a short distance before being comminuted by chemical weathering and abrasion (Büdel, 1982, p. 149).

According to Schumm (1981), alluvial channels with large

suspended loads should develop a meandering pattern with a high sinuosity index. Evidence is insufficient to indicate how applicable this relationship is to tropical rivers. Wright *et al.* (1974), in a study of 34 major river systems, found that braided and meandering channels were equally common in all climatic regions. Within the equatorial zone, braided channels were slightly more common in tropical–wet regions while meandering characteristics were more common in savanna areas. Baker (1978) argues that empirical equations used to relate channel patterns to sedimentary characteristics that were developed for temperate and semi-arid rivers are difficult to apply to tropical rivers.

Climatic stability

Landforms that are distinctive to the humid tropics should be found in areas of long-term climatic and tectonic stability. The presence of a thick, often leached, regolith in tropical areas appears to indicate such stability (Douglas, 1969). However, evidence from deep-sea cores and palynological studies shows that the Quaternary period in the tropical zone was one of drastic climatic changes, although interpretations of these changes vary. The evidence for South America, for example, has suggested to some that glacial episodes were tropical–wet regimes (Bé *et al.*, 1976); while others contend that they were cooler and drier than today (van Geel and van der Hammen, 1973). One problem is that climatic fluctuations based on palynological studies of vegetational changes often represent relatively short time-spans. In northern South America during the late Pleistocene and Holocene, the periods of change, which were often cyclic, ranged from 2000 to less than 10000 years in length (van der Hammen, 1974). Peterson *et al.* (1979) contend that climatic regimes varied considerably during the several thousand-year period centred on 18000 years BP. Graham (1976) suggests the possibility that the tropical rain forest of Veracruz, Mexico, may be less than 11000 years old. The equatorial forest is considered to be in a present state of flux and that Quaternary climatic variations are sufficient to prevent any stable equilibrium being attained (Flenley, 1979, p. 128).

Although the geomorphologist must appreciate the probable effect of such climatic changes, it raises the question of the time-span necessary for landform development or landscape modification. The pattern of landscape changes resulting from these periodic fluctuations defy generalisation. Thick regolith records little of such changes, although the weathering profile itself reflects a long period of intense chemical weathering. It is difficult to attribute the formation of such denudational features as planation surfaces to climate changes of this duration. The presence of terrace gravels associated with tropical rivers is explained by discharge characteristics in a semi-arid climate (Vogt, 1959). Tricart (1973, p. 245) relates African examples of alternating mudflows and alluvial sands and valley trenching and aggradation to paleoclimatic oscillations. Baker (1978) suggests that the channel patterns and deposits of the Amazon Basin rivers reflect cooler and drier conditions due to Pleistocene climatic changes. A search for inheritances from a previous climatic regime can lead to circular reasoning because of our lack of knowledge concerning present geomorphic processes.

Tectonic activity

The role of tectonic events in landform evolution is highlighted in Chapter 12; approximately 18 per cent of the tropical rain forest lies within tectonically active zones (Garwood *et al.*, 1979). Currently, active seismic zones are found mainly in South-east Asia and the Americas, although earlier crustal movements must also have influenced Central Africa. Sapper (1935), in a pioneering work on tropical geomorphology, attributed deep, V-shaped tropical valleys to vertical erosion by streams. Although many of his observations and examples were taken from Central America, he did not consider tectonic activity as a major cause of valley incision. The idea that incision and valley formation is a major geomorphic process in the tropical rain forest, unrelated to tectonic activity, is still prevalent today. Löffler (1977) notes that the alluvial plain adjacent to the Central Mountain Range in New Guinea, covered by wet–tropical forest, is much more dissected than its distal margin covered by savanna and monsoon forest. He attributes this difference to erosional processes within the tropical forest and considers that tectonic forces have had little impact.

Tropical alluvial fans

Clearly, there are two particular problems for students of tropical geomorphology. First, it is difficult to accept the argument that landform processes in the tropics are totally different from the mid-latitude environment in the absence of sufficient, detailed information (Büdel, 1982, p. 121). Secondly, the best-known denudational landforms in the tropics take significant periods of geological time to develop, during which both climatic and tectonic changes can be comparatively abrupt and substantial. Therefore, the present chapter now considers a specific, dynamic area in Central America in some detail, and focuses on a particular, but little-studied depositional landform. Although little studied in the tropical zone, detailed studies in arid and semi-arid areas (Bull, 1977) have shown that alluvial fans adjust rapidly to environmental changes as well as reflecting more persistent controls.

General description of the General Valley and the Rio Toro Amarillo

A series of coalescing, inactive alluvial fans were studied in the Rio General Valley in south-eastern Costa Rica along with an active fan, the Rio Toro Amarillo, on the Atlantic watershed (Figure 7.1). The General Valley is situated between two parallel mountain ranges; the Cordillera de Talamanca, along the north-eastern margin, which reaches an elevation of 3820 m; and the Coast Range, along the south-eastern margin, with a maximum elevation slightly above 1300 m. The Talamanca is composed largely of Tertiary marine sediments with interstratified basaltic and andesitic rocks and intrusive rocks composed of quartz diorite and granodiorite (Weyl, 1980, p. 146). The Coast Range is made up of mid- to late Tertiary marine sedimentary rocks and Pliocene volcanic rocks. Volcanic activity, although significant during the Pliocene, did not occur in this portion of Costa Rica during the Quaternary (Weyl, 1980, p. 230). The structural trough in which the General Valley is located was formed by differential uplift of the bordering mountain ranges during late Pliocene or early Pleistocene time and continuing into the Holocene (Weyl, 1980, p. 149). Following uplift, the General Valley was floored by alluvial fans that extend out 10-16 km from the base of the Talamanca and are derived from erosion of that range (Kesel, 1983). With former uplift of the bounding mountain ranges, the main fan-forming streams incised their channels below the adjacent fan surface, extending from the apex to the distal end. Sediment transport and deposition is now confined to the floor of the incised valleys. The upper fan surfaces no longer receive sediments from the source basin and can be considered inactive. These upper fan surfaces range in age from late

Figure 7.1: Location of alluvial fans, Costa Rica.

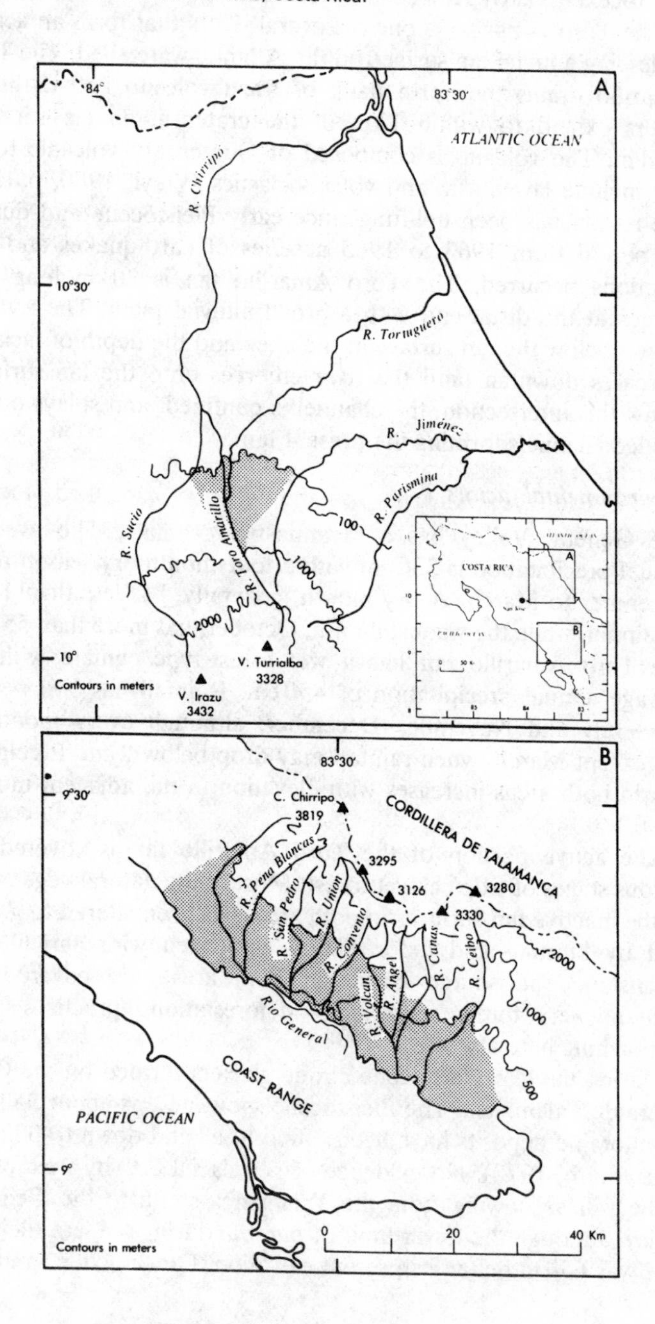

Pleistocene to early Holocene (Kesel, 1983).

The Toro Amarillo is one of several rivers that form an active, coalescing alluvial fan surface on the Atlantic watershed. The Toro Amarillo drains the north flank of Irazu volcano and its headwaters extend to within 1 km of the crater at an elevation of 3364 m. The volcano is composed of Quaternary volcanic rocks that include lavas, ash, and volcanoclastics (Weyl, 1980, p. 131). The region has been uplifting since early Pleistocene and during the period from 1963 to 1965 a series of earthquakes and ash eruptions occurred. The Toro Amarillo fan is 20 km long and merges at the distal end with a broad alluvial plain. The river is incised below the fan surface at the apex and the depth of incision decreases downfan until the river emerges onto the fan surface. Below this intersection, the channel is confined, and splays out in braided channels forming the present fan.

Environmental factors

The General Valley has a savanna-type climate. The average annual precipitation is 330 cm with a four-month dry season from December to March. A dry month, generally, has less than 1 cm of rainfall, while the wettest month, October, has more than 55 cm. The Toro Amarillo fan has a wet forest-type climate with an average annual precipitation of 450 cm. Rainfall has two peaks, June–July and November–December, although every month is wet except March, when rainfall may drop below 2 cm. Precipitation in both areas increases with elevation in the adjacent mountains.

The active portion of the Toro Amarillo fan is covered by various stages of tropical wet forest. Most of the natural vegetation on the inactive fan surfaces in both areas has been altered to grassland by burning and deforestation associated with agricultural expansion. The source basins in both areas are covered by montane-wet forest, although deforestation practices are encroaching here also.

During the late Pleistocene cirque glaciers formed on the Cordillera de Talamanca. The Pleistocene snowline was about 3500 m and moraine deposits have been found to extend down to 3320 m (Hastenrath, 1973). No evidence of periglacial activity was found in the valleys flowing from the Talamanca south to the General Valley, although the lower limit of *paramo* during the late Pleistocene was 650 m below its present elevation. Other pollen samples from the alluvial fan sediments in the General Valley suggest that climatic conditions in the valley from 17 000 years BP to the present fluctuated between grass savanna with scattered woodland and a period with a slight increase in precipitation represented by increase in oaks (Kesel, 1983). The present savanna climate dates from about 9000 years BP.

Fan morphology

The relationship between the area of the Costa Rican fans and their source basin is compared with a wide range of other examples (Figure 7.2). The Jamaican fan (Yallahs) is not a true fan (Wescott and Ethridge, 1980), but a fan delta and therefore might be expected to plot closer to the Puerto Rican fan (Salinas), which has a similar size source basin (McClymonds and Ward, 1966). The Costa Rican, Honduran, and Indian fans are located in highly active tectonic zones (Gole and Chitale, 1966; Schramm, 1981). The high sediment yields produced from arid and semi-arid basins and low yields from humid basins (Langbein and Schumm, 1958) might suggest that tropical fan areas would be less than arid fans. In fact, the Costa Rican fans are larger than many arid fans with comparable size source basins (Figure 7.2). They are similar in size to fans formed in other tectonically active areas like the Elburz Mountains (Beaumont, 1972), or where the source basin is composed of highly erodible rock types (Bull, 1964). The positions of the Kosi River and Honduran fans coincide with the projected trend line of the Costa Rican fans.

Studies of the gradient of the fan surface have ranged from using the overall longitudinal profile (Bull, 1964) to the upper fourth of the fan surface where the gradient should be most controlled by the characteristics of the source basin upstream (Melton, 1965). The latter was used here because the number of perennial channels that influenced the fan surface profile increases downfan. Factors influencing fan gradient include source-basin areas, discharge, mode of sediment deposition, and rate of sediment production in the source basin (Bull, 1977). A comparison of fans from various climatic areas with similar size source basins (Figure 7.3) indicates that arid fans have a much steeper gradient than glacial outwash rivers which have been used as analogous to tropical fan rivers (Boothroyd and Nummedal, 1978). The gradients of the Costa Rican fans, however, are as steep as the arid fans which suggests that sediment characteristics and tectonic events play a

Figure 7.2: Relation of fan area to basin area.

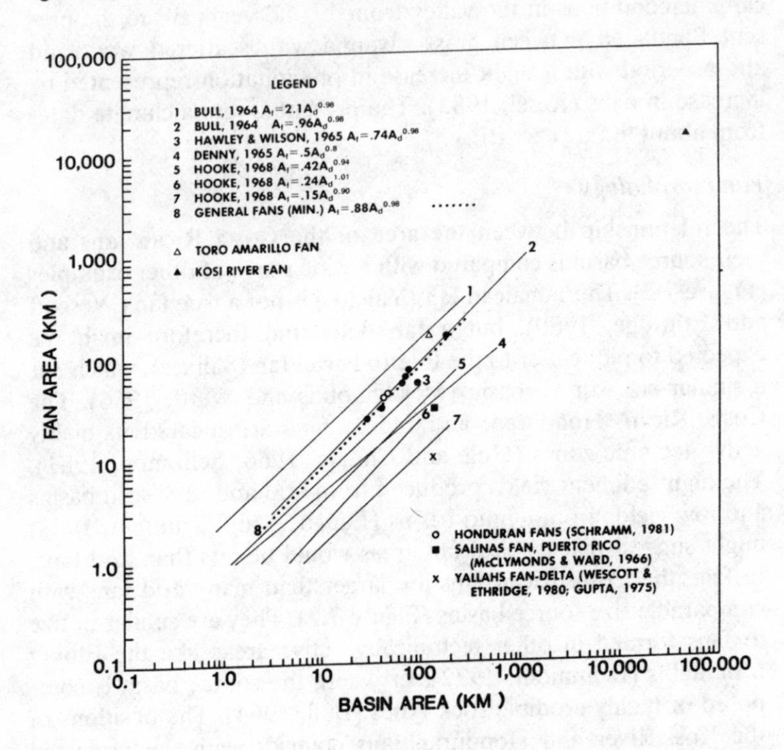

much greater role than climate-related factors in determining fan gradient.

Depositional pattern

The Toro Amarillo cross-fan profile is similar to the arid fans described by Bull (1964), with a marked convexity in the proximal portion that flattens toward the distal margin. Stream channels at the apex are located within a 30° arc from the medial position, indicating the range of channel migration on the upper fan. The active fan is being deposited downslope from the intersection point by the constantly shifting channel of the river (Figures 7.4 and 7.5). The river is constantly building a portion of the alluvial cone higher than the adjacent fan surface. Eventually, the river is diverted into the lower areas, and the cycle repeated. Three major zones can be identified on the 1960 photograph (Figure 7.4). The eastern margin is the highest zone and probably represents the for-

Figure 7.3: Longitudinal profiles of alluvial fans forming under different environments.

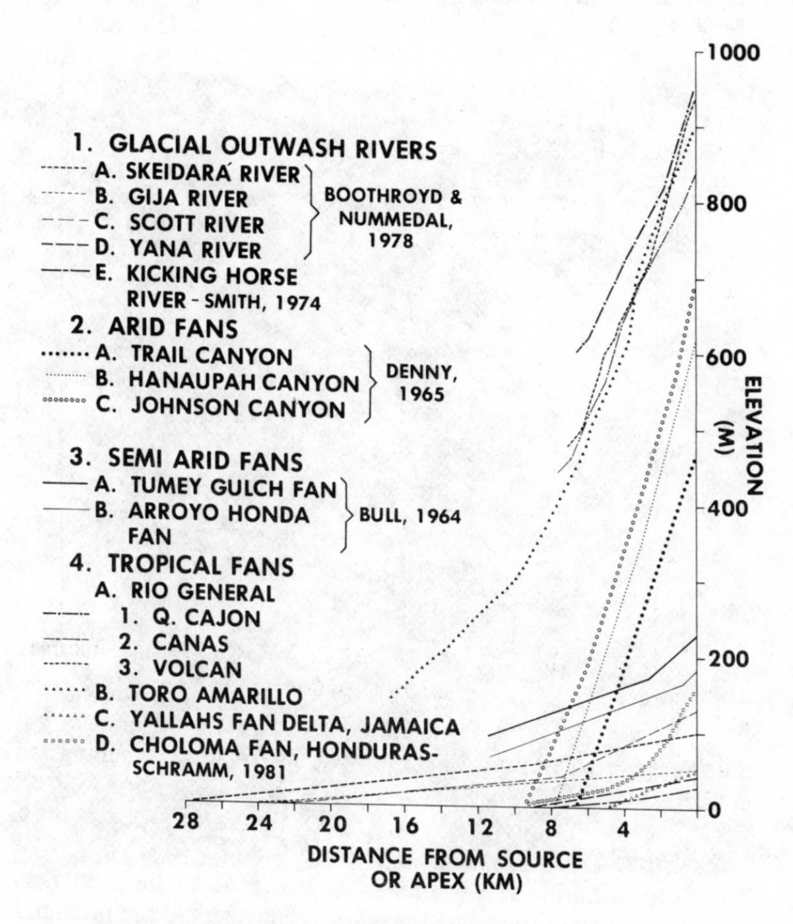

mer channel prior to 1960. This portion is well vegetated and has a single channel that probably carried water only during flood stage. The main channel occupies the central position of the fan and is slightly lower in elevation that the area to the East. The western margin is the lowest zone, and is occupied by another parallel river system which has a tributary that is also a tributary of the Toro Amarillo. The tributary is apparently eroding its channel, thereby increasing the relief between the central alluvial cone and the adjacent western margin. In the 1972 photograph the eastern margin is completely vegetated and no channel is visible (Figure 7.5). A

Figure 7.4: Aerial photograph Toro Amarillo fan, 1960. North is to the top. Runway in south-east corner is 1.1 km long. (Photograph by Instituto Geographico de Costa Rica, San José).

sizeable portion of the main channel flow has been diverted from the central channel into the lower western margin by avulsion. By 1981, most of the flow had been diverted to the western margin and the central position was vegetated, except for a single channel.

The progressive shifting character of the Toro Amarillo channel is similar to that described by Gole and Chitale (1966) for the Kosi River in India. They suggest that the rate of channel shifting is dependent on the rate of deposition on the active alluvial cone. This continuously shifting pattern contrasts with the sudden and often erratic changes in channel position noted on arid fans (Beaty,

Figure 7.5: Aerial photograph, Toro Amarillo fan, 1972. Scale and orientation the same as Figure 7.4. (Photograph by Instituto Geographico de Costa Rica, San José).

1970). There is also evidence on the General Valley fans that sudden shifts in channel position take place as the result of stream piracy when streams that occupy the lower interfan area erode headward on the steeper gradient of the proximal fan margin. Fan head diversions of this type may be associated with humid-type fans, especially in tectonically active areas. Whether piracy was involved in the shifting of the Toro Amarillo is not known.

Episodic sediment regimes and channel patterns

Both the General Valley fans and the Toro Amarillo fan have channel patterns that reflect changes in the amount of sediment supplied. A comparison of the 1960 and 1972 aerial photographs (Figures 7.4 and 7.5) of the Toro Amarillo fan illustrate strikingly the response by the fluvial system to episodic variations in sediment discharge. From 1963 to 1965 a series of ash-producing earthquakes and eruptions occurred on Irazu volcano. Although much of this ash was deposited on the opposite flank of Irazu (Kesel, 1973), sizeable amounts of debris were produced in the Toro Amarillo basin, perhaps partly by earthquakes. This is reflected in the 1972 photograph (Figure 7.5). Previously, eruptions had occurred in 1939-40 followed by an earthquake in 1952 (Murata *et al.*, 1966).

In 1960, the Toro Amarillo followed a somewhat sinuous course in the incised upper part of the fan. The channel has numerous braid bars (Smith, 1974) or islands that are mostly former meander-type point bars and channel side bars that have been cut through by chute and/or transverse channels. They represent depositional bars whose shapes have been modified by erosion. The braid bars are forested and apparently escape flooding. In this reach, flow is confined to a single, well-defined channel that is approximately 50 m wide, although there are sections where the width may reach 100 m during flood. During flood periods, chute and transverse channels may experience water flow. In general, this portion of the 1960 channel appears to be transporting sediment remobilised from the adjacent fan surface.

Where the river emerges onto the active fan surface, there is one dominant channel, that within 2 km divides into several smaller channels. The channel pattern over the entire active fan resembles that of the Kosi River. There is a general increase in the number of channels downfan, but at the distal margin the channels unite and at the merger with the alluvial plain there is a single, somewhat sinuous, well-defined channel. The 1960 channel pattern is better described as anastomosing rather than braided. In the anastomosing pattern the river divides into several sub-channels that are separated by an alluvial surface marked by a dry multi-channel network. Often one of the channels is more dominant than the others. Except for active channels, most of this alluvial surface including braid bars in vegetated. During flood periods, some of the dry-channel network has water flow.

The 1972 Toro Amarillo channel changed to a braided pattern, caused by an increase in sediment discharge. In the upper incised section these changes include an increase in the number of sub-channels; an increase in the width of the active braided channel network; and a marked decrease in the number and size of vegetated braid bars. On the active fan surface similar changes have taken place. The 1972 photograph illustrates the shift of the main channel to a lower interfan position along the western margin and the complete abandonment of the most easterly channel. The numerous braid bars in the 1972 westernmost channel, although depositional in character, have shapes that are completely modified by erosion of existing alluvial deposits. The increase in channel width has also remobilised sediment from the proximal and distal fan that is transported further downfan or to the alluvial plain.

In the General Valley fans, the channel pattern has been influenced by uplift at both the distal and proximal ends. The present drainage pattern consists of single incised channels, or multiple incised channel networks that resemble the 1960 Toro Amarillo pattern. Channel banks and braid bars, where present, are vegetated and there is little evidence of overbank flooding or deposition. A braided channel network that would be representative of an active-fan surface is absent. The present channel pattern appears to reflect a period when sediment supply available from the source basin is low. The last period of major sediment production is yet unknown. In the absence of volcanic activity in this part of Costa Rica during the Quaternary, any episodic production of sediment from the source basins may be related to major earthquakes, or possibly to factors associated with extreme climatic events or changes.

Maximum clast size relationships

Sediments on humid fans are deposited entirely by fluvial processes and strong interrelationships exist between the textural trends of the sediments, fan gradient, and progressive distance downfan (Davis, 1983, p. 236). The tropical fan sediments comprising the Costa Rican fans are dominated by cobble and boulder-size clasts. The clast size variability within the coarse sediments was examined on both the active and the inactive fans using measurements of the intermediate axis (b-axis) of the ten largest clasts. The composition of the largest clasts on the General fans is granodiorite and on the Toro Amarillo fans is various volcanic rocks. The clasts of plutonic rock on the General fans indicate sediment transport from the Cordillera de Talamanca, which is the only pos-

Figure 7.6: Relation of maximum clast size to distance from apex or source.

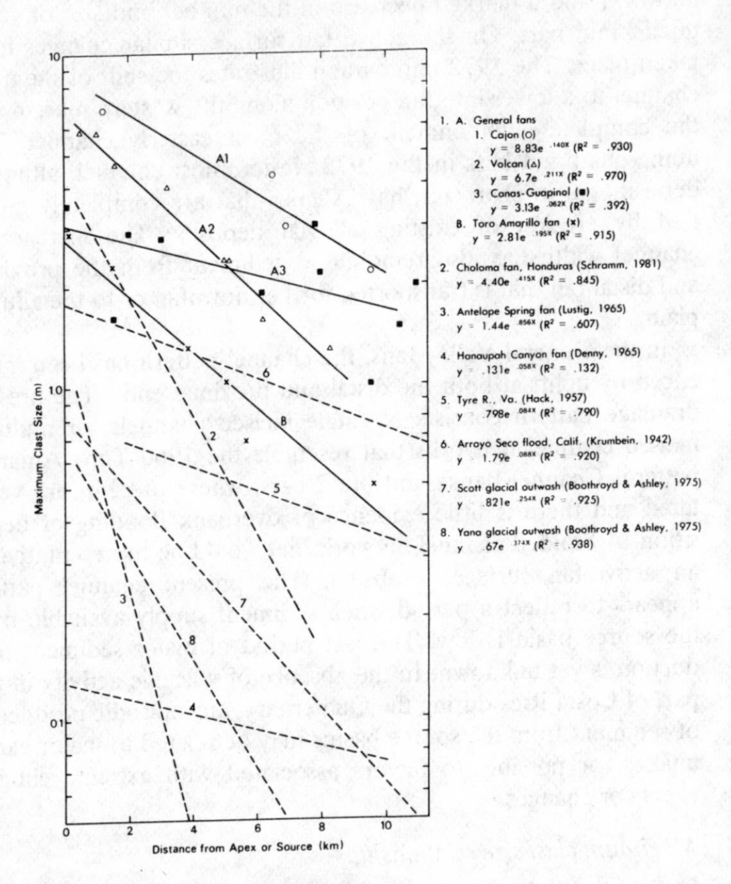

sible source for this rock type (Weyl, 1980, p. 66).

The maximum clast size on the Costa Rican fans decreases in the downfan direction (Figure 7.6), according to fan gradient (Figure 7.7.). Both relationships strongly suggest transport by a low-viscosity medium, probably dominated by fluvial processes. Some of the variability in downfan changes in clast size may be attributed to fan incision, which allows remobilisation of coarse material that is then transported farther downfan (Schumm, 1977, p. 261). Uplift of the proximal fan surface may also play a role by increasing fan gradient. The Toro Amarillo fan has a lower gradient than the General fans (Figure 7.7). This is partly a function of the smaller-sized debris transported by the Toro Amarillo, but

Figure 7.7: Relation of maximum clast size to fan slope for selected Costa Rican fans.

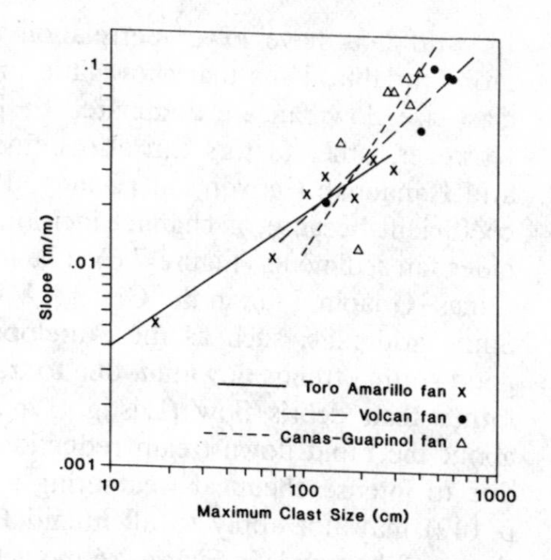

Figure 7.8: Relation of maximum clast size to fan slope for different environments.

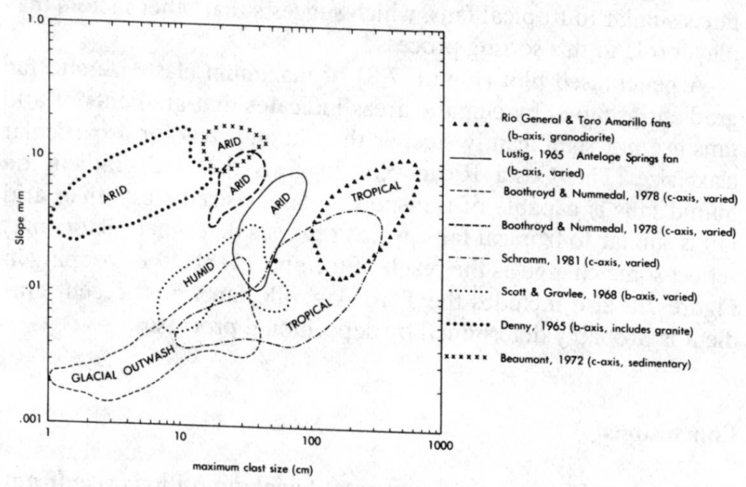

may also reflect greater uplift of the proximal portion of the General fans.

Depositional processes strongly affect the gradient and degree of sorting on fans (Hooke, 1968; Boothroyd and Nummedal, 1978). The Costa Rican fan data compares interestingly with gravel-dominated, braided channels from fans of similar size and lithology. The maximum clast size on tropical fans tends to decrease exponentially downfan (Figure 7.6). Similar relationships for arid fans have lower correlation coefficients and are much more variable. Fans that show little or no decrease in maximum clast size downfan are dominated by high-viscosity debris flows. However, other factors can also influence this relationship. The arid Hanaupah Canyon fan (Denny, 1965) has a low correlation coefficient because of channel incision and the remobilisation of older fan sediments (Figure 7.6). The low correlation value for the Canas–Guapinol fan in the General Valley may reflect this factor. Some arid fans, such as the Antelope Canyon, show generally good sorting trends downfan, due to transport by fluvial processes rather than debris flow (Lustig, 1965). Clearly, generalisations about the rapid downstream reduction of coarse sediment clasts due to intense chemical weathering and abrasion (Büdel, 1982, p. 149) may not apply to all humid tropical environments. The slopes of the trends in Figure 7.6 indicate that the downstream rate of change in clast size in glacial outwash rivers and on one arid fan is similar to tropical fans, which suggests that other factors may play a role in this sorting process.

A generalised plot (Figure 7.8) of maximum clast size and fan gradient grouped by climatic areas indicates that gradients of arid fans are not significantly steeper than humid fans for a particular clast size. The Costa Rican plot suggests that the gradient on humid fans is capable of transporting larger-sized clasts than arid fan is similar to tropical fans, which suggests that other factors may reflect some change as the result of tectonic events, the grouping in Figure 7.8 also includes the Toro Amarillo fan whose recent gradient is probably determined by depositional processes.

Conclusions

Alluvial fans in Costa Rica illustrate fluvial depositional landform development in a tropical environment, within a tectonically active setting. The Toro Amarillo observations suggest that sediment supply to alluvial fans derived from tropical wet basins is episodic in nature. Short phases of active fan building are followed by longer inactive periods. During the inactive phase, little new sediment is supplied to the fan and remobilised older fan sediment is transported to lower portions of the fan. The episodic nature of the sediment supply reflects tectonic and volcanic activity, although abrupt climatic events may also produce this type of response. For the General fan, the longer period cycles related to climatic changes are a possibility.

Channel and depositional patterns on tropical fans are controlled by the continuous shifting of the river course and the episodic character of the sediment supply. A multi-channel braided system characterises fans during periods of high sediment discharge, and an anastomosing pattern, dominated by a single channel, characterises periods of low sediment discharge. The rapid growth rate of tropical vegetation plays a major role in the narrowing of the active channel and the formation of braid bars as the anastomosing pattern develops.

The area of the tropical fans is similar to fans associated with tectonically active zones or easily eroded rock types in extratropical areas. Tectonic activity and sediment size appear to be more important in determining the gradient of fans than are climatic factors. The maximum clast size on the fans exhibits a decrease in size downfan as well as a definite relationship to fan gradient. These relationships strongly suggest transport by fluvial processes. Any decrease in clast size downfan, dependent on comminution by abrasion and chemical weathering, is not evident. A comparison with rivers in other environments suggests that different factors, such as progressive sorting and changes in flow capacity, may be significant. The clast data also indicate that channel slopes on tropical fans are capable of transporting larger-sized clasts than fans with similar gradients in less humid environments.

The tropical fan data imply that tectonic activity can produce fans with similar morphological characteristics regardless of climate. Clearly, further studies of tropical fluvial systems must consider the geological and climatic characteristics, and the tectonic history of the area. The analysis of these systems requires much greater attention to geomorphic processes before any delineation of landforms, distinctive to tropical environments, can be made.

References

Baker, V.R. (1978), 'Adjustments of fluvial systems to climate and source terrain in tropical and subtropical environments', in Miall, A.D. (ed.), *Fluvial Sedimentology* (*Can. Soc. Petrol. Geol.*), pp. 211-30.

Bé, A.W.H., Damuth, J.E., Lott, L. and Free, R. (1976), 'Late Quaternary climatic record in western equatorial Atlantic sediments', in Cline, R.M. and Hayes, J.D. (eds), *Investigation of Late Quaternary Paleoceanography and*

Paleoclimatology. (Geol. Soc. Am. Mem.), *145*, 165-200.
Beaty, C.B. (1970), 'Age and estimated rate of accumulation of an alluvial fan. White Mountains, California, USA', *Am. J. Sci.*, *268*, 50-77.
Beaumont, P. (1972), 'Alluvial fans along the foothills of the Elburz Mountains, Iran, *Palaeogeog. Palaeoclim. Palaeoecol.*, *12*, 251-73.
Boothroyd, C.J. and Ashley, G.M. (1975), 'Process, bar morphology, and sedimentary structures on braided outwashfans, northeastern Gulf of Alaska', in Jopling, A.V. and McDonald, B.C. (eds); *Glaciofluvial and Glaciolacustrine Sedimentation, Soc. Econ. Palaeont. Min. Spec. Publ.*, *23*, 193-222.
Boothroyd, J.C. and Nummedal, D. (1978), 'Proglacial braided outwash: A model for humid alluvial-fan deposition', in Miall, A.E. (ed.), *Fluvial Sedimentology, Can. Soc. Petrol. Geol. Mem.*, *5*, 641-68.
Büdel, J. (1957), 'Die "Doppelten Einebnungsflachen" in den feuchten Tropen', *Z. Geomorphol.*, *1*, 201-28.
Büdel, J. (1982), *Climatic Geomorphology*, Trans. by Fischer, L. and Busche, D. (Princeton Univ., New Jersey).
Bull, W.B. (1964), 'Geomorphology of segmented alluvial fans in western Fresno County, California', *US Geol. Surv. Prof. Pap.*, *352–D*, 89-129.
Bull, W.B. (1977), 'The alluvial-fan environment', *Prog. Phys. Geog.*, *1*, 221-70.
Davis, Jr, R.A. (1983), *Depositional Systems, A Genetic Approach to Sedimentary Geology* (Prentice-Hall, New Jersey).
Denny, C.S. (1965), 'Alluvial fans in the Death Valley region, California and Nevada', *US Geol. Surv. Prof. Pap.*, *466*.
Douglas, I. (1969), 'The efficiency of humid tropical denudation systems', *Trans. Inst. Br. Geog.*, *40*, 1-16.
Flenley, J. (1979), *The Equatorial Rain Forest: A Geological History*, (Butterworths, London).
Garwood, N.C., Janos, D.P. and Browkaw, N. (1979), 'Earthquake-caused landslides: A major disturbance to tropical forests', *Science, NY*, *205*, 997-9.
Gole, C.V. and Chitale, S.V. (1966), 'Inland delta building activity of Kosi River', *J. Hydraul. Div. Am. Soc. civ. Engrs.*, *92*, 111-26.
Graham, A. (1976), 'Late Cenozoic evolution of tropical lowland vegetation in Veracruz, Mexico', *Evolution, Lancaster, Pa.*, *29*, 723-35.
Hack, J.T. (1957), 'Studies of longitudinal stream profiles in Virginia and Maryland', *US Geol. Surv. Prof. Pap.*, *294–B*.
Hastenrath, S. (1973), 'On the Pleistocene glaciation of Cordillera de Talamanca, Costa Rica', *Z. Gletscherk. Glazialgeol.*, *9*, 105-21.
Hooke, R. le B. (1968), 'Steady-state relationships on arid-region alluvial fans in closed basins', *Am. J. Sci.*, *266*, 609-29.
Jennings, J.N. (1983), 'Karst landform', *Am. Scient.*, *71*, 578-86.
Kesel, R.H. (1973), Notes on the lahar landforms of Costa Rica, *Z. Geomorphol.*, *18*, 78-91.
Kesel, R.H. (1983), 'Quaternary history of the Rio General Valley, Costa Rica', *Nat. Geog. Soc. Res. Rep.*, *15*, 339-58.
Krumbein, W.C. (1942), 'Flood deposits of Arroyo Seco, Los Angeles County, California, *Bull. geol. Soc. Am.*, *53*, 1355-402.
Langbein, W.B. and Schumm, S.A. (1958), 'Yield of sediment in relation to mean annual precipitation', *Trans. Am. geophys. Un.*, *39*, 1076-84.
Löffler, E. (1977), 'Tropical rainforest and morphogenic stability', *Z. Geomorphol.*, *21*, 21-61.
Louis, H. (1973), 'Fortschritte und Fragwurdigkeiten in neueren Arbeiten zur Analyse fluvialer Landformung besonders in den Tropen', *Z. Geomorphol.*, *17*, 1-42.
Lustig, L.K. (1965), 'Clastic sedimentation in Deep Springs Valley California', *US*

Geol. Surv. Prof. Pap., 352–F, 131-92.
McClymonds, N.E. and Ward, P.E. (1966), 'Hydrologic characteristics of the alluvial fan near Salinas, Puerto Rico', *US Geol. Surv. Prof. Pap., 550–C,* C231-34.
Melton, M.A. (1965), 'The geomorphic and paleoclimatic significance of alluvial deposits in southern Arizona', *J. Geol., 73,* 1-38.
Murata, K.J., Dondoli, C. and Saenz, R. (1966), 'The 1963–65 eruption of Irazu Volcano, Costa Rica (the period of March 1963 to October 1964)', *Bull. Volcan., 29,* 765-96.
Peterson, G.M., Webb, T. III, Kutzbach, J.E., Hammen, T. van der, Wijmstra, T.A. and Street, F.A. (1979), 'The continental record of environmental conditions at 18,000 yrs BP: an initial evaluation, *Quat. Res., 12,* 47-82.
Pippan, T. (1970), 'Characteristics of valley sections in a moderate relief controlled by fluvial erosion (Puerto Rico) compared with such influenced by both fluvial and glacial erosion (Alpine Flysch Zone and Bohemian Forest)', *Z. Geomorphol. Supp., 9,* 119-26.
Sapper, C. (1935), *Geomorphologie der feuchten Tropen* (Leipzig).
Schramm, W.E. (1981), 'Humid tropical alluvial fans, Northwest Honduras, CA', MSc. thesis, Louisiana State Univ.
Schumm, S.A. (1977), *The Fluvial System* (John Wiley, New York).
Schumm, S.A. (1981), 'Evolution and response of the fluvial system, sedimentologic implications', in Ethridge, F.G. and Flores, R.M (eds), *Recent and Ancient Nonmarine Depositional Environments: Models for Exploration, Soc. Econ. Paleont. Spec. Publ., 31,* 31-45.
Scott, K.M. and Gravlee, G.C. (1968), 'Flood surge on the Rubicon River, California — hydrology, hydraulics, and boulder transport', *US Geol. Surv. Prof. Pap., 422–M.*
Smith, N.D. (1974), 'Sedimentology and bar formation in the Upper Kicking Horse River, a braided outwash stream', *J. Geol., 82,* 205-23.
Tricart, J. (1973), *Landforms of the Humid Tropics, Forest, and Savannas,* trans. Kiewiet de Jong, C.J. (St. Martin's Press, New York).
van der Hammen, T. (1974), 'The Pleistocene changes of vegetation and climate in tropical South America', *J. Biogeog., 1,* 3-26.
van Geel, B. and van der Hammen, T. (1973), 'Upper Quaternary vegetational and climatic sequence of the Fuguene area (Eastern Cordillera, Colombia)' *Palaeogeog. Palaeoclim. Palaeoecol., 14,* 9-92.
Vogt, J. (1959), 'Aspects de l'evolution morphologique recente de l'Ouest Africain', *Ann. Geog., 68,* 193-206.
Wescott, W.A. and Ethridge, F.G. (1980), 'Fan delta sedimentology and tectonic setting Yallahs fan delta, Southeast Jamaica', *Bull. Am. Ass. Petrol. Geol., 64,* 374-99.
Weyl, R. (1980), *Geology of Central America,* (Gebruder Borntaeger, Berlin–Stuttgart).
Wright, L.D., Coleman, J.M. and Erickson, M.W. (1974), 'Analysis of major river systems and their deltas: morphologic and process comparisons', *Coastal Stud. Inst., Louisiana State Univ. Tech. Rep., 156.*

8 THEMES IN DESERT GEOMORPHOLOGY

Andrew S. Goudie

Introduction

A prime feature of desert geomorphological research over the past century or so has been the rapidity with which ideas have changed, and the dramatic way in which ideas have gone in and out of fashion. This reflects the fact that hypothesis formulation has often preceded detailed and reliable information on form and process, and the fact that different workers have written about different areas where the relative importance of different processes may vary substantially. For example, scientists working in the Southern Namib, with winds of unimodal direction and extremely high velocity blowing sand for 65 per cent of the time, might have a very different impression from scientists working in a low wind-energy, mountainous terrain, where sheetfloods rather than sand storms might be the dominant activity.

The purpose of this chapter is to investigate the way in which certain key concepts have changed over this period: the role of wind erosion, classification of dune pattern, the effectiveness of desert run-off processes, the causes of rock disintegration, the importance of palaeoclimatic influences, the age of major deserts, the view of deserts obtained from space, and the contribution of applied geomorphology.

Charles R. Keyes and 'extravagent aeolation'

Although the power of wind as a cause of desert landform development was espoused by many French and German geomorphologists working in South-west and North Africa (Passarge, 1904; Walther, 1924), it was an American who most powerfully outlined the role of aeolian activity. The following quotation illustrates his views, which have since been characterised as 'extravagent aeolation' (Cooke and Warren, 1973):

> As in the eighteenth century marine planation was one of the notable discoveries in earth-study, and as in the last century the theory of gradual peneplanation through stream-corrosion was one of the grander conceptions of the age, so the recognition of desert windscour as the principal among erosional agencies seems destined to take its place among the first half-dozen great and novel thoughts which shall especially distinguish geologic science of the twentieth century. (Keyes, 1913, p. 468)

Keyes believed that material weakened by insolation weathering became rock-waste that was rapidly exported through the agency of winds, and deposited as dust sheets on desert margins. The most striking result of such activity was held to be the formation of great plains 'smoother than any peneplain possibly can be'.

> Where but in a dry climate does entire absence of foothills characterize the mountain ranges? Towering desert eminences rise out of elimitable expanse of level plain as volcanic isles jut from the sea. Plain meets mountain as sharply as the strand-line of the ocean.

Keyes' views never received very wide currency and were soon ignored or rejected, particularly in the inter-war years. Several reasons can be postulated for this dismissal. (i) Following Blackwelder's (1933) and Griggs' (1936) experimental work the role of insolation as a cause of rock-waste production was questioned; (ii) the great pediment landscapes of the American South-west were widely attributed to fluvial processes involving sheetflood activity (McGee, 1897); (iii) it became apparent that many of the world's great dust sheets were the product of deflation from glacial areas rather than from deserts (Smalley and Krinsley, 1978); (iv) many desert landscapes were thought to have been moulded by formerly more widespread and powerful 'pluvial' forces; (v) it was widely held that salt and clay crusts, together with lag gravels in the shape of stone pavement, and the presence of a water-table would limit the extent to which aeolian lowering could occur; and (vi) analysis of wind data showed that many deserts were not characterised by especially high velocity winds. Features that were admitted to have an aeolian origin (e.g. pedestal rocks, ventifacts, yardangs) were thought to be but minor embellishments, whilst other possibly aeolian features (e.g. stone pavements and closed depressions) were also explicable by other means. Stone pavements, for example, could be the product of horizontal removal of fine sediment by sheetflood activity (Cooke, 1970); while closed depressions could be attributed to a tectonic, solutional, or bestial origin (Goudie and Thomas, 1984).

One consequence of the wholesale rejection of aeolation was the belief that many desert landscapes were relict features of Pleistocene, or earlier, humid conditions, and that under current prevailing aridity many desert landscapes are largely inactive. This view was typified by Walton's (1969) analysis of the factors shaping desert morphology (pp. 48-9):

> True aridity mummifies a landscape by slowing down the rates of change just as the dry sand and desiccating air preserved Egyptians and Peruvians for the museums of the twentieth century. As these mummies once had vitality, growth and change so had the arid zone landscapes in the past, with but few exceptions. Such landscapes betray the results of higher precipitation and reduced rates of evaporation during the 'pluvials'.

Many of the arguments that were raised against 'extravagant aeolation' were both cogent and sound. None the less, in the last decade or so, it has been realised that there may have been an overreaction to the aeolianist ideas. Certain new lines of evidence which, whilst not seeking to diminish the absolute importance of water action, have brought about a reassessment of wind power.

Cooke (1981) has demonstrated in laboratory simulation and field observations that salt weathering may effectively prepare rock waste for subsequent deflation (Goudie and Watson, 1984).

Analysis of meteorological records, studies of the composition of deep-sea core sediments, and the tracing of dust storm trajectories using satellites, have indicated that in many arid areas deflational dust storm activity occurs frequently (Péwé, 1981). Dust can accumulate on desert margins at an appreciable rate (Goudie, 1983) and is a viable explanation for the development of loess deposits in areas like Bahrain, the Negev, Tunisia and Pakistan.

Palaeoclimatic studies, based on the examination of ocean core sediments, have indicated that dust-storm activity was more widespread during the frequent and severe arid phases of the Pleistocene (Thiede, 1979); that trade wind velocities may have been

Table 8.1: Yardangs, locations and lithology

Location	Lithology
Taklimakan, China	Fluvial and lacustrine sediments of Pleistocene age
Lut. Iran	Pleistocene fine-grained horizontally bedded silty clay and limey gypsiferous sand
Khash Desert, Afghanistan	Clay
Sinai, Egypt	Nubian sandstone
Saudi Arabia	Caliche, limestone
Bahrain	Aeolianite (Pleistocene), dolomite
Egypt	Eocene limestone, lake beds, Nubian sandstone
South-Central Algeria	Cretaceous clays, Cambrian claystones
Borkou, Chad	Palaeozoic and lower Mesozoic sandstones and shales
Jaisalmer, India	Eocene limestones
Namib Desert, Namibia	Precambrian dolomites, granites and gneisses
Rogers Lake, California	Dune sands and lake beds
Northern Peru	Weakly to moderately consolidated upper Eocene to Palaeocene sediments (shales and sandstones)
South-Central Peru	Siltstones of upper Oligocene to Miocene age (Pisco formation)

higher during glacial inter-pluvials (Parkin and Shackleton, 1973); and that many deserts are of considerable antiquity, thereby providing a long period over which aeolian processes might operate.

The study of air photographs, satellite images, and detailed topographic maps has indicated that certain aeolian features that were once regarded as bizarre embellishments or mere curiosities, notably yardangs, are features of great areal extent on a wide range of rock types (Table 8.1).

Desert depressions are extremely widespread (Figure 8.1) in areas like Argentina, North America, South Africa, Botswana and Southern and Western Australia. While many may owe something to non-aeolian processes, their shapes, orientations and associated lunette dunes are powerful indicators of the role of wind. That the basins are not infilled argues that in many areas fluvial processes are ineffective (Goudie and Thomas, 1984).

The pattern of dunes

There is little doubt that sand dunes are areally extensive, and there is no doubt that they are formed predominantly by wind

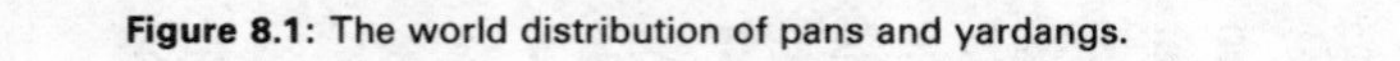
Figure 8.1: The world distribution of pans and yardangs.

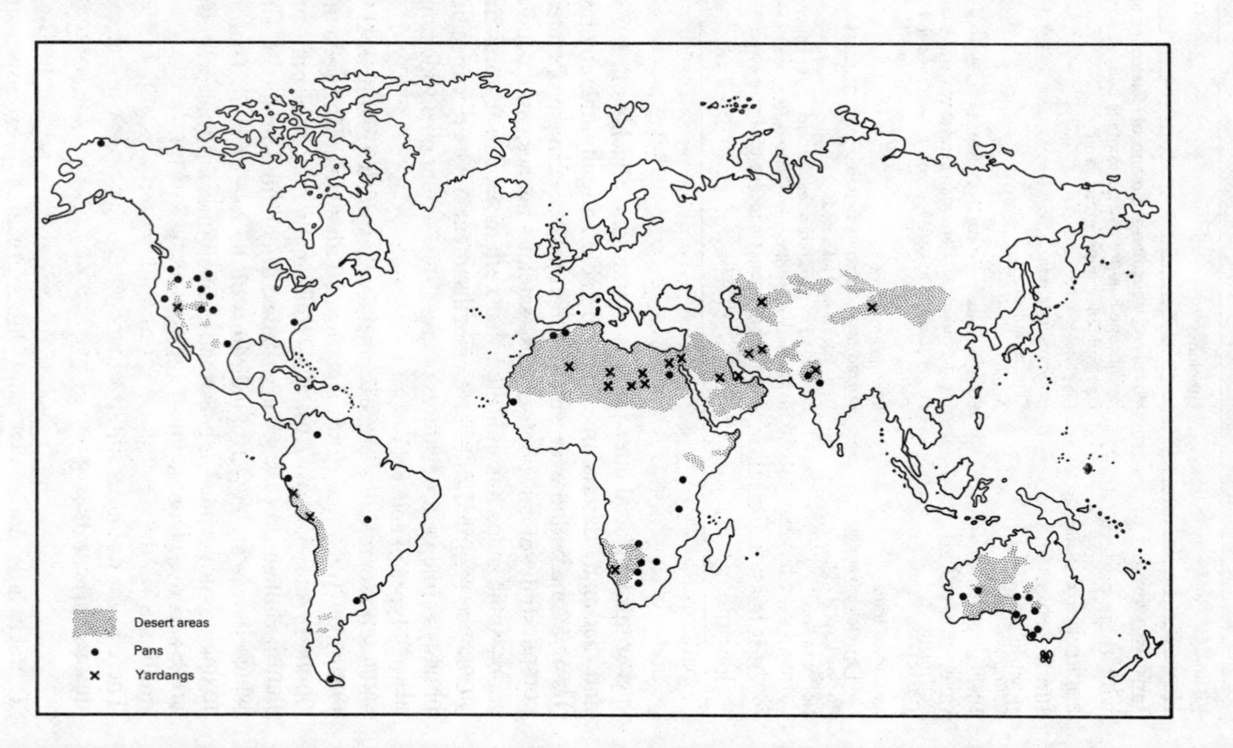

action. Bagnold's classic laboratory and field geomorphological studies of the dynamics of sand movement (Bagnold, 1941) provided fundamental understanding of how and when sand moves. However, he was more concerned with the physical properties of sand movement and with the cross-sectional form of dunes than with their pattern, and cited two simple patterns to illustrate the applicability of some of his principles (p. 189): 'Dunes assume two fundamental shapes: (i) the *barchan* or crescentic dune, and (ii) the longitudinal dune, whose arabic name *seif* or "sword" will be used hereafter for brevity.' Simple barchans are, in reality, rather rare, and it is now normal to stress the complexity of dune patterns. As Cooke and Warren (1973, p. 285) state:

> Dune patterns are not simple in nature. The huge number of aerial photographs of arid areas that have recently become available have revealed that complicated patterns are the rule. . . . It has been common to describe dunes as either transverse or longitudinal, but one of the most obvious properties of dune patterns is that they fail to conform to these simple classifications: oblique forms are not only common, they are virtually universal.

Thus most classification schemes for sand dunes are relatively complicated, as indicated in Figure 8.2.

McKee (1979) has identified certain basic forms — sheets, stringers, domes, barchans, barchanoid ridges, transverse ridges, blowouts, parabolics, linears (seifs), reversing and stars. There are also compound dune types which consist of two or more of the same type combined by overlapping or by being superimposed on one another (e.g. star dunes coalescing to form an intricate pattern or arms and peaks, or little barchan dunes resting on the windward flanks of large ones). Finally, there are complex dune types in which two basic types of different form have grown or coalesced together (e.g. linear dunes in parallel rows with star dunes on their crests). Our understanding of the conditions which give rise to this multitude of different dune forms and patterns is still limited. There is, for example, still controversy as to the types of wind conditions that give rise to one of Bagnold's basic types, the seif dune.

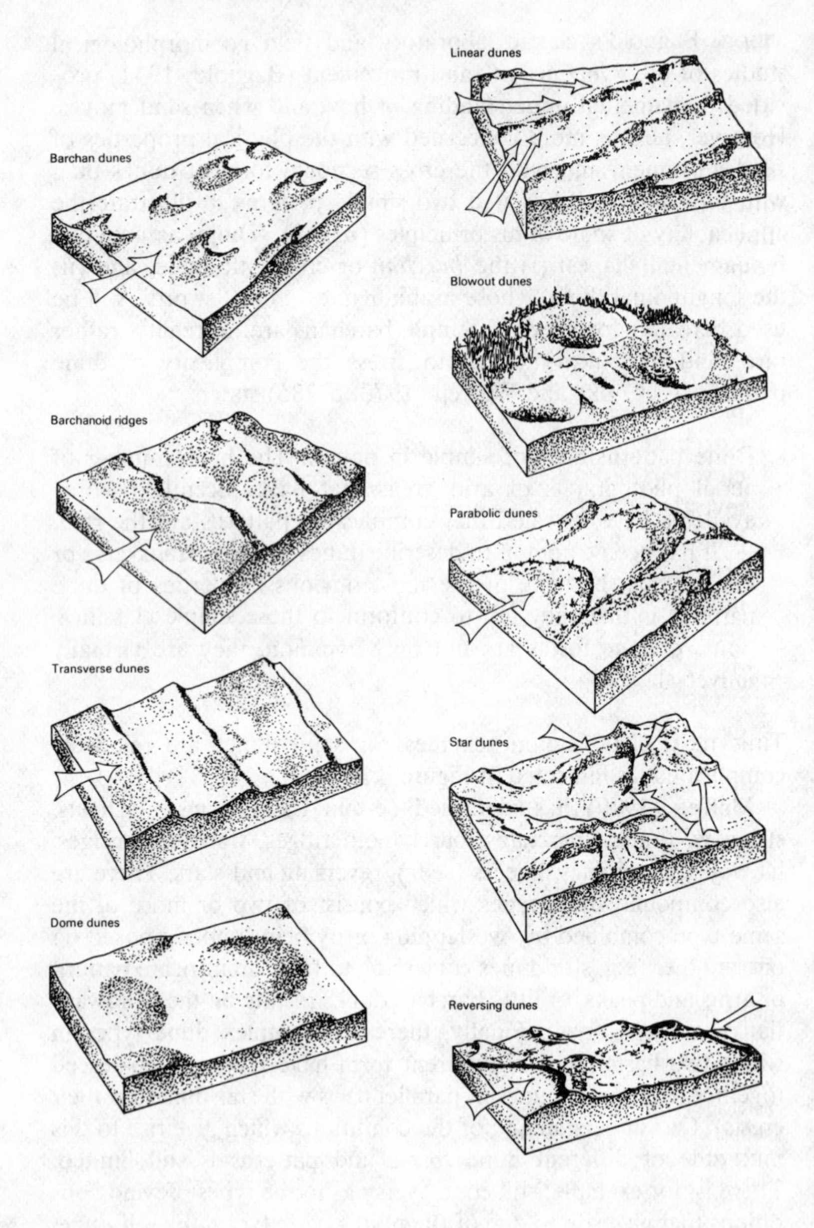
Linear dunes
Barchan dunes
Blowout dunes
Barchanoid ridges
Parabolic dunes
Transverse dunes
Star dunes
Dome dunes
Reversing dunes

The power and role of water

There are many desert areas where running water is the dominant process. Indeed, even certain major aeolian processes — dust-storm generation and sand provision for erg development — may often depend on the provision of a source of silt or sand from fluvial deposits. However, as Peel (1975, p. 110) has noted with respect to ideas that were current earlier in the century:

> That rainfall and streamflow could have made any significant contribution seems scarcely to have occurred to many of these pioneers, for these were phenomena seldom encountered in the extreme Old World deserts. Their introduction, and progressive elevation into the leading role, came mainly with the geological investigation of the American Far West; for in that huge region lofty rain and snow-catching ranges interweave with the semi-arid plateaux and markedly arid sunken basins in a varied and intricate network. In the mountains powerful rivers could be seen sawing out canyons, and seasonal floods continuing the dissection of the plateaux, while similar floods poured down occasionally even into the most desiccated basins. Almost everywhere the imprint of water-action was clearly visible.

Despite extremely high rates of evaporation, combined with large areas being blanketed by non-cohesive sand and gravel, large

Figure 8.2: E.D. McKee and his collaborators (McKee, 1979) have used satellite imagery to map the different types of dune found in the world's major deserts. The major classes that they encountered are shown here, together with their formative wind directions (shown by the arrows). Barchans are crescentic-shaped individual dunes. The horns of the crescent point downwind. Barchanoid ridges are asymmetric waves, oriented transverse to the wind direction, which consist of coalesced barchans in rows. They have steeply dipping lee slope. Transverse dunes also form normal to the dominant wind direction but lack barchanoid structures. Dome dunes form where dune height is inhibited by unobstructed strong winds. They lack a steep slip face. Linear dunes are parallel, straight dunes with slip faces on both sides and with their length many times greater than their width. Blowouts result from erosion of a pre-existing sand sheet, while parabolics represent a type of blowout in which the middle part has moved forward with respect to the sides or arms. In areas of complex winds, dunes may have a starlike form while, in areas with winds from two opposing directions, reversing dunes may form. In addition to these simple forms, more complex patterns combine various elements.

fractions of desert surfaces display patterns of recent waterflow, and devastating floods are a notorious feature of certain desert areas. Individual storm totals produced by frontal systems or small convective cells can occasionally be very high, and while intensities can also reach 1 mm/min, it would be misleading to exaggerate the rarity and intensity of desert storms. For instance, the average storm in Death Valley, California, and in the centre of the Jordanian Desert, only tends to produce about 4 mm of rainfall, and around 20 rainy days may occur during the course of a year in areas where the mean annual rainfall is less than 100 mm. It seems that desert floods are as much the result of the nature of desert surfaces as they are the result of the special nature of desert rainfall. Quite small rainfall events generate run-off (Table 8.2). Factors responsible include the lack of interception by vegetation, the poor structure of many surface materials, and the tendency for soils to become mechanically crusted by rainsplash. The run-off that is generated in deserts helps to account for the formation of a wide range of striking landforms found in arid areas, including arroyos (Graf, 1983), alluvial fans (Bull, 1977), canyons, and the concave profiles of many rock-cut pediments (Rahn, 1967).

The disintegration question

Angular rock debris, split boulders, tafoni and alveoles, and sheeting structures on bare rock outcrops are among the many phenomena that have prompted geomorphologists to question the causes

Table 8.2: Data on the intensity of rainfall generating run-off in arid areas

Area	Intensity generating run-off	Total (mm)
Mojave, California	34 mm/24hr	—
Negev, Israel	1.7 mm/3 min	18
	3.4 mm/10 min	—
Ahaggar, Sahara	25.4 mm/10 min	—
Central Australia		12.7–50.8
Nahel Yael, Israel	1.5 mm/3 min	8–30
Walnut Gulch, USA	—	< 6
Alice Springs, Australia	44.6 mm/50 hr	—

Source: From Cooke *et al.* (1982), Table VI.7, based on various sources.

of rock breakdown in deserts. Such breakdown is self-evident, but the causes continued to be debated. As Schattner (1958, p. 248) has related:

> Up to the thirties of the present century granular disintegration, scaling-off, exfoliation and to some extent the formation of certain types of alveoli, particularly 'tafoni', were thought to be the result of mechanical or physical weathering and specifically characteristic of dry desert conditions. It was assumed that they were primarily a result of insolation, which builds up on the bare rock surfaces minute but ever-recurring stresses due to high temperatures during the day and contractional stresses due to the sharp drop of temperature at night.

The arguments used in support of the insolation hypothesis included the great diurnal ranges of temperature occurring under cloudless skies, reports of rocks going bang in the night, the existence of large amounts of angular rock debris with little sign of chemical alteration, and certain rock characteristics that might facilitate disintegration through expansion and contraction — polymineralcy, polychromacy, and disorder in texture. However, as the twentieth century progressed, doubts began to arise as to the effectiveness of this mechanism. One of the prime reasons was laboratory simulation work (Tarr, 1915; Blackwelder, 1933; Griggs, 1936). Laboratory rock specimens, even though they were subjected to a higher range of temperatures than those endured by rocks in their natural environment, suffered little or no change in the absence of water. But in the presence of water some change did occur. In addition, studies of rock disintegration in natural environments where moisture was present — as near the River Nile — showed that breakdown was more prevalent than in drier situations. Furthermore, in the Mojave Desert, Roth (1965) found that even the interiors of boulders in desert regions may contain appreciable moisture, derived from atmospheric humidity, dew, fog and occasional precipitation. Another setback to the insolation hypothesis was that sheeting structures were found on rock exposures in other environments (e.g. the bornhardts of savanna and rainforest areas). Similarly, spheroidal weathering was frequently found to occur on rocks unexposed to the sun and lying under a debris cover many metres in depth. For reasons such as these, the power of insolation weathering was more or less dismissed, and other mechanisms, especially hydration or pressure release, were put forward. The whole contribution of mechanical weathering processes, which with wind action had once been held to be one of the diagnostic and characteristic geomorphological processes of the arid realm, was relegated to secondary importance.

Since the mid-1960s, however, the power of physical weathering processes has been reassessed. There are three main reasons for this: doubts have been expressed about the validity of some of the earlier experimental work; the theory behind insolation weathering has been re-examined; and other mechanical weathering mechanisms have been shown to be capable of substantial rock breakdown. The early experiments were carried out on small, unrestrained cubes, made of a limited range of rock types. The weathering cycles were not representative of natural conditions. More seriously, polished specimens tended to be used and this may have reduced spalling tendencies. Furthermore, the methods used to detect micro-cracking were not as sophisticated as those used today (Aires-Barros *et al.*, 1975). The most telling criticism, however, is based on the small cube size employed, for the larger the specimen, the larger the thermal stresses will tend to be: 'A 10-cm-diam rock has 10 times the spalling tendency of a 1-cm-diam pebble' (Rice, 1976, p. 61).

Of the two alternative mechanical weathering mechanisms that have been proposed as being important in deserts, the hydration mechanism is one for which there is still little empirical evidence. Some rocks, notably argillaceous dolomites and shales, may fail as a result of fatigue induced by the expansion and contraction of absorbed water during repeated cycles of wetting and drying (Dunn and Hudec, 1972). Likewise, Ravina and Zaslavsky (1974) have theorised that stresses can be induced in rocks by the interaction of electrical double layers on adjacent mineral surfaces. The other mechanical weathering mechanism, which involves crystal growth and/or hydration of salts, has both a sound theoretical basis (Evans, 1969) and abundant empirical evidence for its operation. Field observations (Goudie and Watson, 1984), laboratory simulations (Sperling and Cooke, 1980), engineering studies (Jones, 1980), and appreciation of the widespread field occurrence of salt, have shown very clearly that salt weathering is a potent geomorphological mechanism in many desert areas.

The palaeoclimatic problem

The degree to which desert landscapes are the product of past pluvial phases has been much debated. The extreme view of Büdel (1982) on this matter was based on various assumptions about the nature of Tertiary climates. Writing on arid zones (p. 201) he said:

> The most salient geomorphological feature of this zone is the major relief assemblage ... inherited from pre-Pleistocene humid periods.... About 75 % of this relief has survived the dry periods which dominated most of the Quaternary. Only the remaining 25 % or so has undergone traditional planation or even transformation.

Büdel believed that desert landscapes were dominated by humid-tropical etchplains, and that 'fully desert conditions can only have been widespread in the Sahara since the Plio-Pleistocene transition' (p. 204). The Pleistocene was regarded as too short to have been a period of major landscape modification. There is some support for his viewpoint in the form of widespread duricrusts associated with deep weathering profiles in areas like Central Australia (Langford-Smith, 1978). On the other hand, it is uncertain if major desert areas in the trade wind belt are a product of post-Pliocene times, or if Tertiary climates were uniformly warm and humid. For example, although the rate of dust output from deserts into the oceans has accelerated during Tertiary and Quaternary times (Janackek and Rea, 1983), aeolian components have recently been found in Cretaceous and Tertiary North Atlantic sediments. Clays and quartz have been supplied to the North Atlantic by wind blowing from the American and African continents since the early Cretaceous. Furthermore, Lever and McCave (1983) believe that the zone of maximum input has remained in palaeolatitude 20-30° N since the early Cretaceous. The cool upwelling waters of the Senguela current were promoting aridity along the Namib Coast of South-west Africa (Namibia) by the late Miocene, although N. Lancaster believes that arid conditions prevailed in the Namib during the mid-Eocene to Upper Miocene. He also believes that the inland Kalahari has an equally long record of arid conditions, and may have existed well back into the Mesozoic. Other deserts which appear to be of considerable antiquity are those of China, for evidence of aridity has been traced back to the late Cretaceous and early Tertiary (Chao and Xing, 1982).

Another concept which is related to the belief that many desert landscapes are relict landscapes is that of post-glacial progressive desiccation (Goudie, 1972). For long it was believed that the Pleistocene was characterised by pluvial conditions, and that true aridity was a product of the Holocene. However, a great discovery of the last few decades is that, over much of the lower latitudes, glacial phases tended to be relatively dry, that present inter-glacial conditions have only been characteristic of a relatively small proportion of the Quaternary, and that large parts of the Tropics have been subjected to greatly increased areas of sand desert (erg). The imprint of such formerly more extensive deserts is revealed by the presence of degraded dunes and other palaeo-aeolian forms in areas like the High Plains of North America (Wells, 1983), the Kalahari (Grove, 1969; Thomas, 1984), the Mato Grosso of South America (Klammer, 1982), the Thar (Allchin *et al.*, 1978), and Australia.

Most desert landscapes preserve a highly complex record of frequent and severe climatic changes, and while ancient river systems, old archaeological sites, fossil karst phenomena, high lake strandlines, and deep weathering profiles provide clear evidence of more humid conditions, there is also abundant evidence for the existence of many phases when increased aridity also occurred.

Deserts from and in space

One of the most interesting recent developments in arid geomorphology has resulted from studies of the planet Mars, which has abundant aeolian features. Water action is also clearly evident on its surface. Bars, braids and dendritic networks have been attributed to fluvial action, and probable catastrophic flood channels of 'channelled scabland type' have been reported (Baker, 1978). Telescopic observations from earth have revealed the presence of dust clouds on Mars for a considerable time, and it is now known that they can be large enough to shroud the entire planet (Mills, 1977) and last for up to four months at a time. Silt and clay-sized particles appear to be widespread, and possibly as a result of electrostatic forces (increased in their intensity of operation because of the low atmospheric pressure conditions) they may be agglomerated to form sand-sized aggregates analogous to some of the clay dunes known from earth. Indeed, low atmospheric pressure has many important geomorphological consequences, particularly in terms of threshold velocities for sand movement. It has been estimated that the threshold-drag-velocity required for a 200 μ grain might be as much as 10 times greater on Mars than on earth. Saltating grains should therefore have 10 times greater momentum and 100 times more kinetic energy than those on earth (McCauley, 1973). Furthermore, atmospheric cushioning would also be almost non-existent, thus permitting particles in the very small size ranges to act as efficient instruments of abrasion.

Saltating at levels 3 to 4 times higher than on earth, and carried by high velocity winds, they smash into bedrock, other grains and other objects with great ferocity and are fragmented into tiny particles. Experiments simulating Martian conditions show that sand particles can be reduced to less than 20 μ in diameter in about 20 minutes. One consequence of this heightened sand movement activity is the development of wind erosional features, including streamlined ridges, modified crater rims, irregular pits and hollows, linear grooves, fluted cliffs, and reticulate ridges (McCauley, 1973). Sand movement has also created dune patterns on Mars, and Breed (1977) has suggested that the crescentic ridges of the Hellespontus dunes have comparable morphometry to dunes in the Karakum and the Ala Shan deserts on earth. Enormous barchans also occur, with dimensions of around 100 × 300 km (Belcher *et al.*, 1971).

Desert landform studies on earth have benefited from the search for analogues for features on Mars (El-Baz and Maxwell, 1982), and certain fundamental studies on sand movement and experimental abrasion have added to the theoretical basis of terrestrial studies. The Western Desert of Egypt appears to be the closest analogue to Mars, and among features that have been identified in both locations are sharply defined inselbergs, boulder tracks, sand streaks, pitted boulder fields and canyonlands.

Space research has benefited desert geomorphology most directly by providing a mega-scale approach to morphological and distributional studies. The Landsat programme, for example, has enabled the first comprehensive maps to be prepared of world dune types (McKee, 1979), and has enabled such global patterns to be related to regional wind characteristics. Satellite imagery has also been used in the evaluation of dune trends and distribution to infer palaeoclimate (Ahlbrandt and Fryberger, 1980). Analysis of

Table 8.3: Relative importance of major dune types in the world's deserts

	Thar	Takia Makan	Namib	Kalahari	Saudi Arabia	Ala Shan	South Sahara	North Sahara	North-east Sahara	West Sahara	Average
A. Linear dunes (total)	13.96	22.12	32.55	85.55	49.81	1.44	24.08	22.84	17.01	35.49	30.54
simple and compound	13.96	18.91	18.50	85.85	26.24	1.44	24.08	5.74	2.41	35.49	23.26
feathered	—	—	—	—	4.36	—	—	3.56	1.13	—	0.91
with crescentic superimposed	—	3.21	—	—	—	—	—	4.02	7.32	—	1.46
with stars superimposed	—	—	14.34	—	19.21	—	—	9.52	6.15	—	4.92
B. Crescentic (total)	54.29	36.91	11.80	0.59	14.91	27.01	28.37	33.34	14.53	19.17	24.09
single barchanoid ridges	8.96	3.21	11.80	—	0.59	8.62	4.08	0.06	—	0.65	3.80
megabarchans	—	—	—	—	—	—	—	7.18	1.98	—	0.92
complex barchanoid ridges	16.65	33.70	—	—	14.32	18.39	24.29	26.10	12.55	18.52	16.45
parabolics	28.68	—	—	0.59	—	—	—	—	—	—	2.93
C. Star dunes	—	—	9.92	—	5.34	2.87	—	7.92	23.92	—	5.00
D. Dome dunes	—	7.40	—	—	—	0.86	—	—	0.80	—	0.90
E. Sheets and streaks	31.75	33.56	45.44	13.56	23.24	67.82	47.54	35.92	39.25	45.34	38.34
F. Undifferentiated	—	—	—	—	6.71	—	—	—	4.50	—	1.12

Source: Analysis of maps of Breed *et al.* (1979) by A.S. Goudie

the Landsat-derived maps by Breed *et al.* (1979) enables an assessment of the relative importance of the major dune types in the world's deserts (Table 8.3). Planimetric studies suggest on a world basis that the most common form of aeolian depositional surface is that of sand sheets and streaks (*c.* 38 per cent). Second in importance are miscellaneous types of linear dune (*c.* 30 per cent), followed by crescentic dunes of predominantly barchanoid type (*c.* 24 per cent), star dunes (5 per cent) and dome dunes (*c.* 1 per cent). Major regional differences arise between ergs. The Kalahari is notable for having a predominance of linear dunes (*c.* 86 per cent), the Thar for having parabolics (*c.* 29 per cent), and the North-east Sahara for having star dunes (*c.* 24 per cent). Satellites have also facilitated a better appreciation of the source areas, trajectories and frequencies of dust storms and the distribution and alignment of yardangs.

The application of geomorphological research

One of the most striking features of desert research is its long-standing concern with matters that have social, economic or engineering significance. There are two prime reasons for this: the need to find and develop resources, and the need to avoid natural and quasi-natural hazards. Resource development in arid lands has for long occupied geomorphologists. For example, the pioneer surveys carried out in the American West by the US Geological Survey spawned such remarkable geomorphologists as J.W. Powell, W.J. McGee and G.K. Gilbert. Although they made major contributions to the initial development of theory in geomorphology, and in the 1880s coined the term itself, their initial role was to pursue environmental surveys that would assist in the development of the frontier states. More recently, geomorphologists have been

Table 8.4: Examples of geomorphological hazards in deserts

Flooding of valleys, fans and playas	Sedimentation behind dams	Calcretization
Sabkha inundation	Clear water erosion downstream from dams	Desiccation contraction
Hydrocompaction	Salt weathering	Dune encroachment and reactivation
	Gully development	
	Piping	
Surface subsidence due to water and hydrocarbon extraction	Landslides, rock falls and related slope failure phenomena	Visibility reduction by dust
		Soil erosion

employed widely in the interior of Australia by CSIRO and other bodies to create resource inventories. With the development of the arid lands of the Middle East in the 1970s and 1980s there has been a need to locate aggregate resources to supply the construction industry (Doornkamp *et al.*, 1980). In the field of natural hazard assessment, the role of the applied geomorphologist has been most important (Cooke *et al.*, 1982). Arid areas are characterised by a very wide range of geomorphological and hydrological hazards (Table 8.4). Thus the Dust Bowl years in the USA in the 1930s prompted fundamental studies of soil movement by wind (Chepil, 1946); the collapse of buildings because of salt attack in the Middle East caused a reappraisal of the role of salt weathering (Cooke, 1981); the problems of flooding in Israel encouraged detailed and original studies of fluvial processes (Yair *et al.*, 1978); and sand-encroachment problems have generated a substantial body of research on aeolian dynamics. Furthermore, geomorphologists have contributed to the whole question of desertification (Warren and Maizels, 1977), and have considered some of the many ways in which man may have created or exacerbated some of the problems of desert areas. Significantly, experience in some of the fragile and degraded lands of the Middle East persuaded Marsh (1864) to develop some of his perceptive ideas on the role of man in changing the face of the earth and to initiate calls for the conservation of nature.

References

Ahlbrandt, T.D. and Fryberger, S.G. (1980), 'Eolian deposits in the Nebraska sand hills', *US Geol. Surv. Prof. Pap.*, *1120A*.

Aires-Barros, L., Graca, R.C. and Velez, A. (1975), 'Dry and wet laboratory tests and thermal fatigue of rocks', *Engng. Geol.*, *9*, 249-65.

Allchin, B., Goudie, A.S. and Hegde, K.T.M. (1978), *The Prehistory and Palaeo-geography of the Great Indian Desert* (Academic Press, London).

Bagnold, R.A. (1941), *The Physics of Blown Sand and Desert Dunes* (Methuen, London).

Baker, V.R. (1978), 'The Spokane flood controversy and the Martian outflow channels', *Science, NY.*, *202*, 1249-56.

Belcher, D., Vererka, J. and Sagan, C. (1971), 'Mariner photography of Mars and aerial photography of earth: some analogues', *Icarus*, *15*, 241-52.

Blackwelder, E. (1933), 'The insolation hypothesis of rock weathering', *Am. J. Sci.*, *26*, 97-113.

Breed, C.S. (1977), 'Terrestrial analogs of the Hellespontus dunes, Mars', *Icarus*, *20*, 326-40.

Breed, C.S. *et al.* (1979), 'Regional studies of sand seas, using Landsat (ERTS) imagery', *US Geol. Surv. Prof. Pap., 1052*, 305-97.
Büdel, J. (1982), *Climatic Geomorphology* (Princeton Univ. Press, Princeton).
Bull, W.B. (1977), 'The alluvial fan environment', *Prog. Phys. Geog., 1*, 222-70.
Chao, S.C. and Xing, J.M. (1982) 'Origin and development of the Shamo (sandy deserts) and the Gobi (stony deserts) of China', *Striae, 17*, 79-91.
Chepil, W.S. (1946), 'Dynamics of wind erosion: IV. The translocating and abrasive action of the wind', *Soil Sci., 61*, 167-77.
Cooke, R.U. (1970), 'Stone pavements in deserts', *Ann. Ass. Am. Geog., 60*, 560-77.
Cooke, R.U. (1981), 'Salt weathering in deserts', *Proc. Geol. Ass., 92*, 1-16.
Cooke, R.U. and Warren, A. (1973), *Geomorphology in Deserts* (Batsford, London).
Cooke, R.U., Brunsden, D., Doornkamp, J.C. and Jones, D.K.C. (1982), *Urban Geomorphology in Drylands* (Oxford University Press, Oxford).
Doornkamp, J.C., Brunsden, D. and Jones, D.K.C. (1980), *Geology, Geomorphology and Pedology of Bahrain* (Geobooks, Norwich).
Dunn, J.R. and Hudec, P.P. (1972), 'Frost adsorption effects in argillaceous rocks', *Highw. Res. Rec., 393*, 65-78.
El-Baz, F. and Maxwell, T.A. (eds) (1982), *Desert Landforms of South-west Egypt: A Basis for Comparison with Mars* (NASA, Washington).
Evans, I.S. (1969), 'Salt crystallization and rock weathering: a review', *Rév. Géomorphol. Dyn., 19*, 153-77.
Goudie, A.S. (1972), 'The concept of post-glacial progressive desiccation', *Univ. Oxford. Sch. Geog. Res. Pap.*
Goudie, A.S. (1983), 'Dust storms in space and time', *Prog. Phys. Geog., 7*, 502-30.
Goudie, A.S. and Thomas, D.S.G. (1984), 'Pans in Southern Africa with particular reference to South Africa and Zimbabwe'. *Z. Geomorphol.* (in press)
Goudie, A.S. and Watson, A. (1984), 'Rock block monitoring of rapid salt weathering in southern Tunisia', *Earth Surf. Proc. Landf., 9*, 95-8.
Graf, W.L. (1983) 'The Arroyo problem — palaeohydrology and palaeo-hydraulics in the short term', in Gregory, K.J.(ed.), *Background to Palaeohydrology* (Wiley, Chichester), pp. 279-302.
Griggs, D.T. (1936), 'The factor of fatigue in rock exfoliation', *J. Geol., 44*, 783-96.
Grove, A.T. (1969), 'Landforms and climatic change in the Kalahari and Ngamiland', *Geog. J., 135*, 192-212.
Janackek, T.R. and Rea, D.K. (1983) 'Eolian deposition in the North-east Pacific Ocean: Cenozoic history of atmospheric circulation', *Bull. Geol. Soc. Am., 94*, 730-78.
Jones, D.K.C. (1980), 'British applied geomorphology: an appraisal', *Z. Geomorphol. Supp., 36*, 48-73.
Keyes, C.R. (1913), 'Great erosional work of winds', *Pop. Sci. Mon.*, May, 468-77.
Klammer, G. (1982) 'Die palaeowüste des Pantannal von Mato Grosso und die Pleistozone klimageschichte der Brasilianaschen Randtropen', *Z. Geomorphol., 26*, 393-416.
Lancaster, N. (1982), 'Linear dunes', *Prog. Phys. Geog., 6*, 475-504.
Langford-Smith, T. (ed.) (1978), *Silcrete in Australia* (Dept. Geog. Univ. New England, Armidale).
Lever, A. and McCave, I.N. (1983), 'Eolian components in Cretaceous and Tertiary North Atlantic sediments', *J. Sedim. Petrol., 53*, 811-32.
McCauley, J.F. (1973), 'Mariner 9 evidence for wind erosion in the equatorial and mid-latitude regions of Mars', *J. Geophys. Res., 78*, 4123-38.

McCauley, J.F., Breed, C.S. and Grolier, M.J. (1982), 'The interplay of fluvial mass-wasting and eolian processes in the eastern Gilf-Kebris region', in El-Baz, F. and Maxwell, T.A. (eds), 207-40.

McGee, W.J. (1897), 'Sheetflood erosion', *Bull. Geol. Soc. Am.*, *8*, 87-112.

McKee, E.D. (ed.) (1979), 'A study of global sand seas', *US Geol. Surv. Prof. Pap.*, *1052*.

Marsh, G.P. (1864), *Man and Nature.*

Mills, A.A. (1977), 'Dust clouds and their generation of glow discharges on Mars', *Nature, Lond.*, *268*, 614.

Parkin, D.W. and Shackleton, N.J. (1973), 'Trade winds and temperature correlations down a deep-sea core off the Saharan coast', *Nature, Lond.*, *245*, 455-7.

Passarge, S. (1904), 'Die inselberglandschaften in tropische Afrika', *Naturw. Wschr.*, *3*, 657-65.

Peel, R.F. (1975), 'Water action in desert landscapes', in Peel, R.E., Chisholm, M. and Haggett, P. (eds), *Processes in Physical and Human Geography* (Heinemann Educational Books, London), pp. 110-29.

Péwé, T.L. (ed.) (1981), 'Desert dust: origin, characteristics and effects on man', *Geol. Soc. Am. Spec. Pap.*, *186*.

Rahn, P. (1967), 'Sheetfloods, streamfloods, and the formation of pediments', *Ann. Ass. Am. Geog.*, *57*, 593-604.

Ravina, I. and Zavslavsky, D. (1974), 'The electrical double layer as a possible factor in desert weathering', *Z. Geomorphol. Supp. 21*, 13-18.

Rice, A. (1976), 'Insolation warmed over', *Geology*, *4*, 61-2.

Roth, E.S. (1965), 'Temperature and water content as factors in desert weathering', *J. Geol.*, *73*, 454-68.

Schattner, I. (1958), 'Weathering phenomena in the crystalline rocks of the Sinai in the light of current notions', *Bull. Res. Coun. Israel*, *10 G*, 247-66.

Smalley, I.J. and Krinsley, D.H. (1978), 'Loess deposits associated with deserts', *Catena*, *5*, 53-66.

Sperling, C.H.B. and Cooke, R.U. (1980), 'Salt weathering in arid environments: II. Laboratory studies', *Bedford Coll. Univ. Lond. Pap. Geog.*, *9*.

Tarr, W.A. (1915), 'A study of some heating tests and the light they throw on the disintegration of granite', *Econ. Geol.*, *10*, 348-67.

Thiede, J. (1979), 'Wind regimes over the late Quaternary South-west Pacific Ocean', *Geology*, *7*, 259-62.

Thomas, D.S.G. (1984), 'Ancient ergs of the former arid zones of Zimbabwe, Zambia and Angola', *Trans. Inst. Br. Geog.*, *9*, 75-88.

Walther, J. (1924), *Das Gesetz der Wüstenbildung in Gegenwart und Vorzeit* (von Quelle und Meyer, Leipzig).

Walton, K. (1969), *The Arid Zones* (Hutchinson, London).

Warren, A. and Maizels, J.K. (1977), 'Ecological change and desertification', in United Nations, *Desertification: Its Causes and Consequences* (Pergamon, Oxford), pp. 171-260.

Wells, G.L. (1983), 'Late-glacial circulation over central North America revealed by aeolian features', in Street-Perrott, A., Beran, M. and Ratcliffe, R. (eds), *Variations in the Global Water Budget* (Reidel), pp. 317-30.

Yair, A.D., Sharon, D. and Lavée, H. (1978), 'An instrumented watershed for the study of partial area contribution of runoff in the arid zone', *Z. Geomorphol. Supp.*, *29*, 71-82.

9 ARID ZONE SLOPES AND THEIR ARCHAEOLOGICAL MATERIALS

Ian Reid and Lynne E. Frostick

The revival of interest in process geomorphology at the close of the 1950s brought a sudden increase in information about earth surface processes. This was overdue in the case of hillslope studies where there was increasing frustration with the inability of geomorphology to resolve the controversy surrounding the nature of slope retreat and landscape evolution (Dury, 1959, pp. 59-71). Theoretical considerations had been used to argue both for parallel slope retreat (Penck, 1924) and for the decline of slope angle with age (Davis, 1932). They were also used to explain the morphology of whole continents (King, 1967). Some of these theoretical treatments were described in mathematical terms (Bakker and Le Heux, 1952), or were based on deductive reasoning of the expected physical processes associated with rock comminution and erosion (Wood, 1942). However, there still remained a need to obtain *measured* process rates in order to establish the scale of slope form change that could occur within a given period of time.

The first measurements (Strahler, 1950; Savigear, 1952; Schumm, 1956a; Melton, 1960) clarified some issues but did not remove the controversy. It seemed that slopes could retreat and decline depending upon such factors as the undercutting of a basal stream, and the nature of the weathering mantle. The apparent need for a much larger data-base, together with improvements in the techniques of measurement, led to a significant increase in studies during the 1960s and 1970s (Saunders and Young, 1983). Some of the impetus came from fluvial geomorphologists anxious to locate the sources of river sediments (e.g. Leopold *et al.*, 1966).

Despite this growth of interest in evaluating slope processes, the number of detailed studies from arid environments still remains comparatively small. This is due to the fact that the focus of interest in process geomorphology was largely in humid temperate and sub-tropical latitudes. Visits made by European earth scientists to the world's drylands were of short duration, and so the longer-term assessment of processes was not feasible. Nevertheless, those quantitative studies that were carried out have yielded important results. This is because the substantial rates of erosion on bare ground reduce the significance of the approximations that are inherent in the techniques of measurement.

Besides this, the roles of vegetation and of soil-mixing by soil animals are reduced and the physical processes are clearly visible. As a result, many of these arid zone slope studies have become standard geomorphological examples in textbooks, even though comparatively few in number. Arid zones comprise 36 per cent of the temperate and tropical land surface. Detailed slope studies in these drylands have been conducted in California (Strahler, 1950), Colorado (Schumm, 1964), Alberta (Campbell, 1974), Israel (Yair and Lavee, 1976) and East Africa (Frostick and Reid, 1983), among others. The present account focuses on surface processes operating in the drylands of Northern Kenya. It suggests comparisons and contrasts with other areas, and places the detailed process studies in the broader context of landform evolution.

Northern Kenya

Kenya's Northern Province is semi-arid. Rainfall comes in two short seasons that depend upon the movement of the inter-tropical convergence zone north and south. Annual rainfall is generally less than 400 mm. At the Illeret gauging station close to the experimental site the 11-year record gives an annual average of 308 mm. The prevalence of clear sky conditions over much of the year is reflected in high average temperatures of 30°C. These contribute along with the drying power of strong trade winds, to a high potential rate of evaporation that exceeds 2000 mm a^{-1}. The water balance results in extremely dry soils which can only support a scanty vegetation of thorn scrub. Large areas may be totally devoid of perennial species. The low density of vegetation means a soil with few soil animals; even termite mounds are widely spaced due to the aridity. Limited bioturbation means that soils show little evidence of profile development and contain few biopores. As a result, infiltration rates can be exceedingly low where surficial crusts, formed through rainbeat, are disrupted only by the shrink–swell processes associated with expandible clay minerals.

Hortonian overland flow is more commonly an outcome of

rainfall in this environment than in humid temperate zones. It has been observed to follow the onset of natural rain within 10 minutes (Frostick *et al.*, 1983). This interval compares with that obtained during the sprinkler experiments of Yair *et al.* (1980) in the Negev Desert, Israel where run-off occurred after 5 minutes. The potential for active surface processes is therefore very high. But rates of erosion depend upon the number of rainfalls affecting a particular slope. In fact, there may be no rain for several years, even if an area as a whole is classified as semi-arid and not hyper-arid. This is because rain originates from discrete convective cells that follow erratic paths which may bypass a specific location.

Geologically, there are extensive areas of East Africa underlain by Precambrian basement. Here, inselbergs rise from surrounding pediplains. There are also very large plateaux of basaltic lava flows on which extensive block fields reflect the ready breakdown of ferro-magnesian minerals. Below these plateau surfaces, slope processes and retreat are pronounced in the Plio-Pleistocene sedimentary fills of the great East African Rift basins, especially that of Lake Turkana in Kenya's far north (Figure 9.1). Slopes here can be classified broadly as (1) *Steep straight*: *c.* 30°, cut into sediments that range from silts and clays to gravel conglomerates, and usually associated with scarp valleys. The term talluvium is used to describe the mixture of talus-like fragments and finer colluvium that mantle these slopes; (ii) *shallow straight*: 4-15° , generally cut into silts and silty clays, but with a veneer of coarser lag gravel and fringing the drainage lines on the dissected pediment left by scarp retreat; (iii) *convex straight*: generally developed on clays with an active surface shrink–swell layer of *c.* 10 cm depth. These are similar to Schumm's (1965b) Chadron formation slopes in South Dakota. The processes of types (i) and (ii) were studied, and stereo-paired aerial photographs are provided for visual comparison of each in Figure 9.2.

Surface slope processes

Steep slopes of talluvium

Despite low regional annual rainfall, landform evolution in the Turkana Basin is rapid. The headward extension of drainage networks dissects low scarps of *c.* 100 m with steep-sided valleys that eventually consume the leading edge of the cuesta, resulting in scarp retreat (Figure 9.1. and 9.2). One of these scarp drainage

Figure 9.1: Index maps showing the extent of arid zones on the African continent and details of the drainage network at one of the experimental sites in N. Kenya.

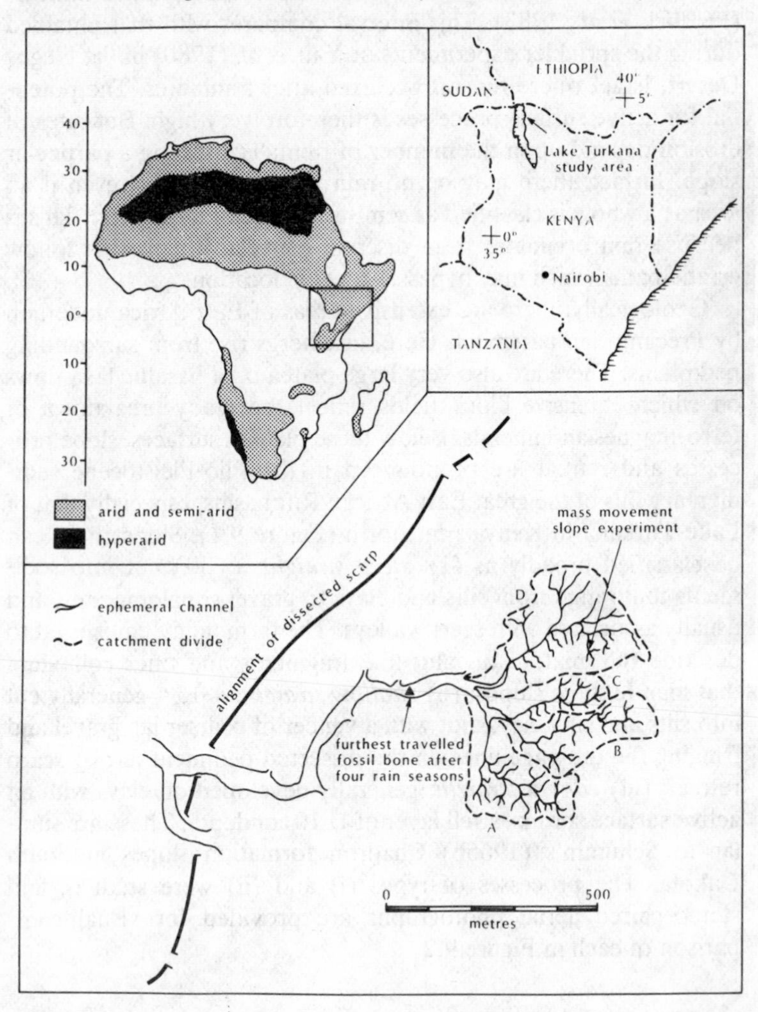

nets has been examined to establish both the rates of displacement of sediment from positions within the slope mantle and its contribution to the stream system.

The scarp is comprised of silts, clays and some sandstone lenses. It is capped by a plateau gravel that varies in thickness between 50 cm and 6 m and which has a near-horizontal surface. Both the gravel and the finer sediments move downslope as a mobile surface

Figure 9.2: *Upper paired photographs:* Typically indented scarp with steep straight 30° slopes clothed with mobile talluvium, N. Kenya. *Lower paired photographs*: typical dissected pediplain left by scarp retreat showing shallow straight slopes of 4-15°. Selective erosion endows the surface of silty soils with a coarse lag gravel.

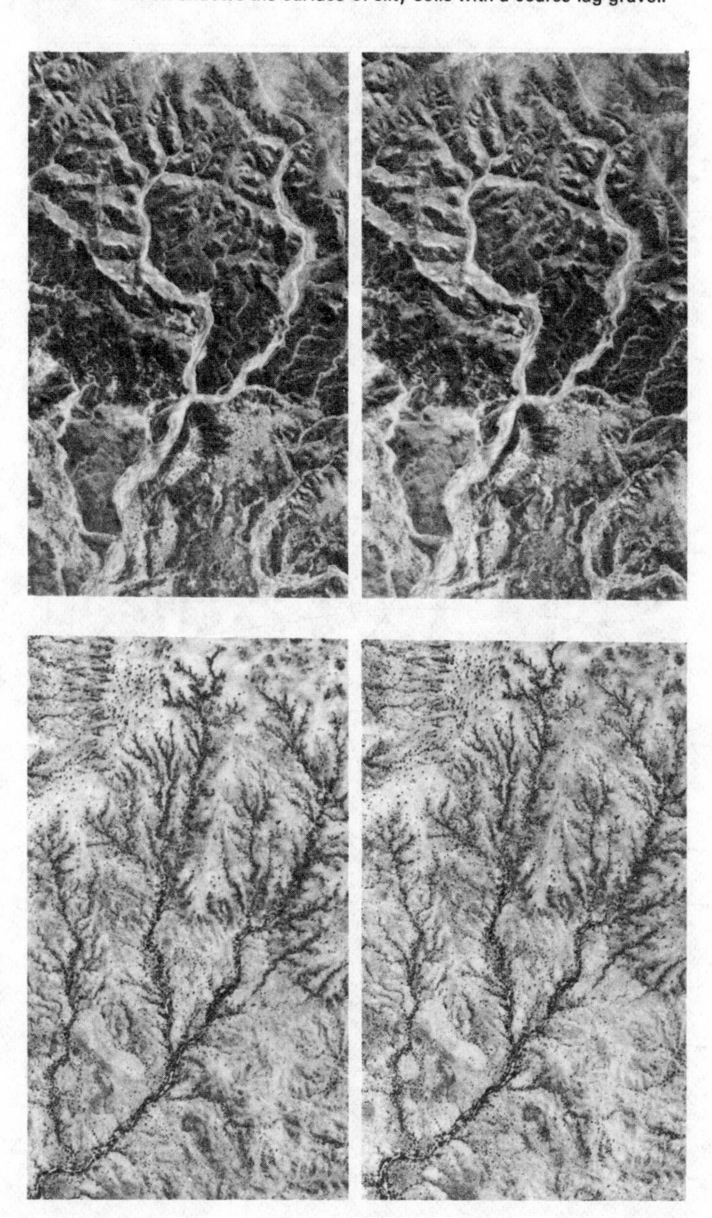

Figure 9.3: Plan of 30° experimental slope showing the effect of two short rain seasons in one year on the displacement of the coarse particles of a talluvial regolith, Koobi Fora Escarpment, N. Kenya.

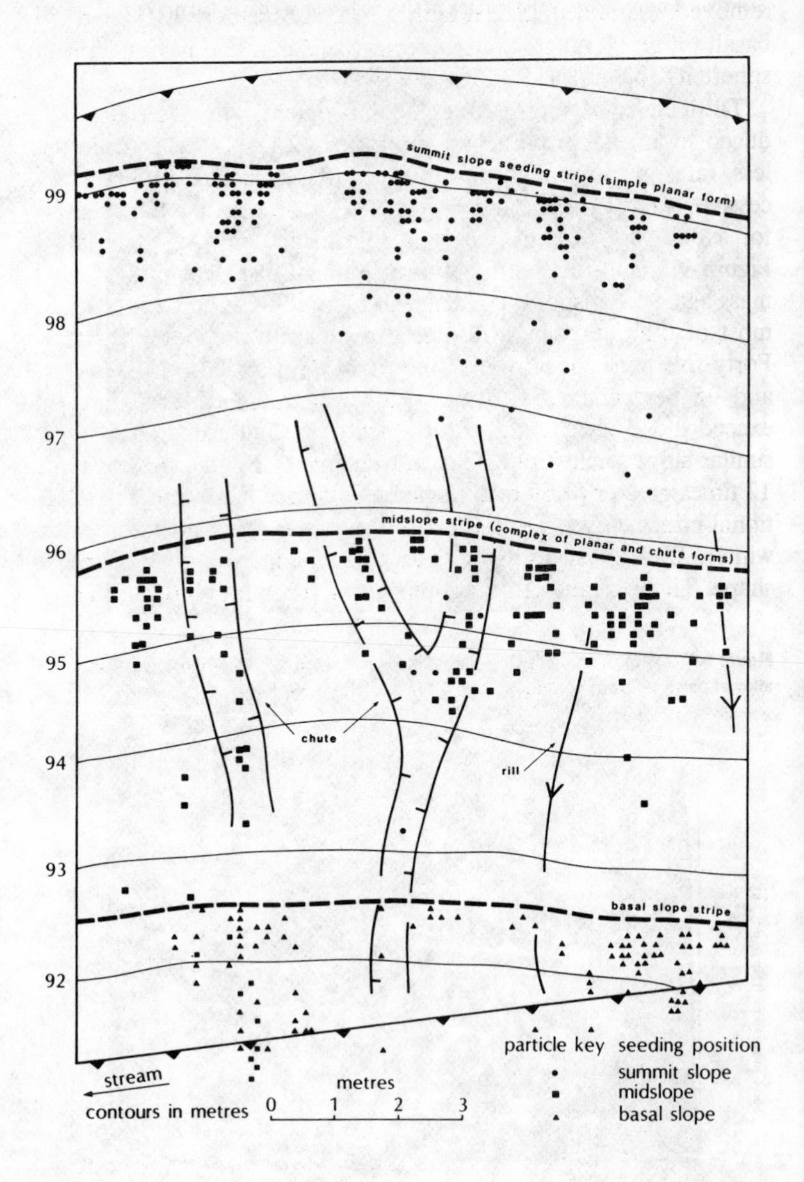

layer of talluvium. On a 30° straight slope, typical of the region, surficial grain movements were assessed over a period of three years. Three contour stripes of coarse particles were carefully removed, painted, replaced and traced. Altogether, 2011 rounded basalt particles, ranging in size from 10-65 mm and having a mean sphericity $[bc/a^2]^{0.33}$ of 0.69 were treated.

Difficulties of interpreting the movements of material introduced to a regolith must be appreciated (Caine, 1981). None the less, rates of movement on the slope were similar in all three successive years (Frostick and Reid, 1982). The plan pattern is given for a one-year (2 rain season, 6 rainstorms) shift of material in Figure 9.3, and the 3-stripe average annual displacement given as mass against transported distance in Figure 9.4. It can be seen that most of the talluvium is displaced comparatively short distances. Forty-five per cent moves < 0.5 m a^{-1}, 65 per cent <1.0 m a^{-1}, and 95 per cent <5.0 m a^{-1}. However, these distances greatly exceed those of Schumm's (1967) study in semi-arid Colorado on similar slope declivities. Mean movements in Northern Kenya are 12 times greater (0.59 m a^{-1} against 0.05 m a^{-1}), despite the additional effect of freeze–thaw processes during the North American winter. This reflects as much as anything the importance of particle shape. The Turkana clasts are more spherical and tend to roll eas-

Figure 9.4: Average annual two-dimensional downslope displacement of coarse alluvial particles from contour stripes 0.1 × 10 m on a 30° slope.

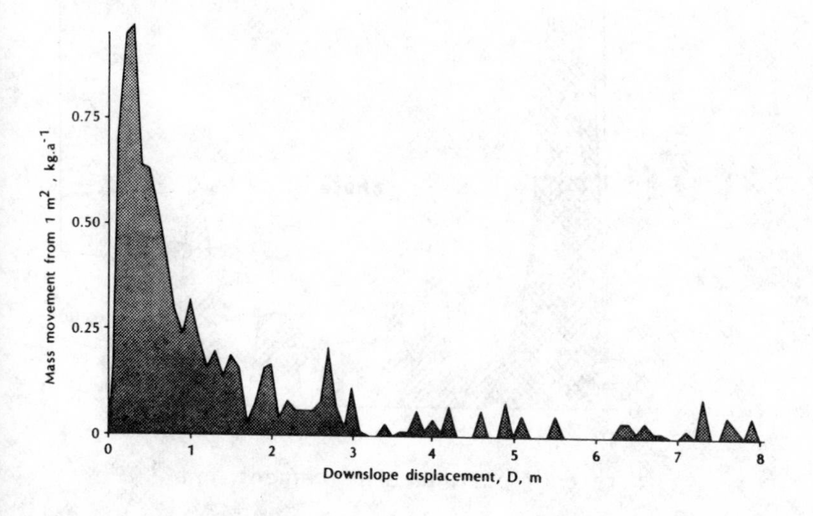

ily once dislodged, whereas the particles in Schumm's study of Colorado slopes are platey and tend to move less far because of the friction involved in sliding.

The actual displacement of surface material may be locally accelerated. If run-off is concentrated so that small contour indentations are created by its increased erosivity, the fine matrix of the talluvium will be etched out, so undermining coarser particles. The transport rate of material in these small chutes is thereby greater than on the major part of the slopes where lack of indentation produces more diffuse run-off (Figures 9.3 and 9.5). In general, the chutes provide the furthest travelled particles that form the right-hand trail of the mass movement portrayed in Figure 9.4.

Shallow wash slopes with lag gravel

Processes on the shallow straight slopes that are characteristic of

Figure 9.5: Frequency distribution of rates of movement of coarse talluvium on simple planar slopes and in small chutes.

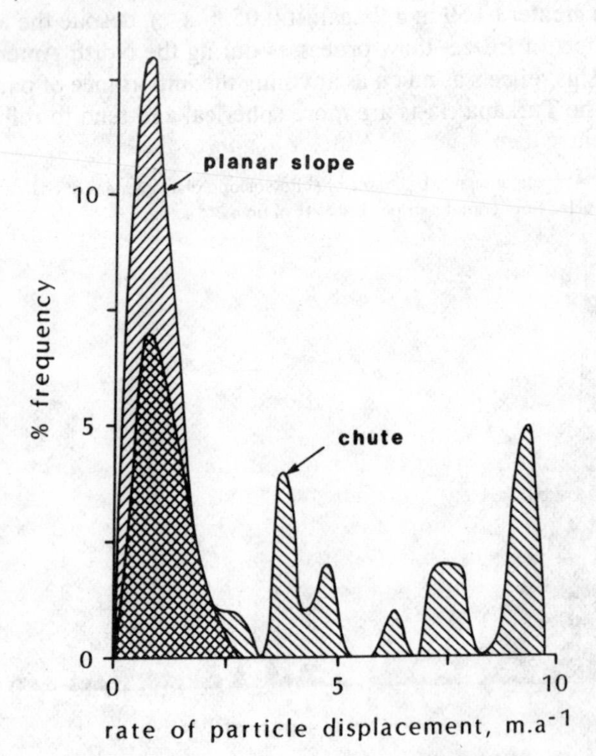

silty-sand and silty-clay lithotypes in many parts of semi-arid Kenya contrast with the gravity-dominated activity of the steep talluvium. The shear force acting on a standard particle lying exposed on a 30° slope is 1.9 times that on a 15° slope, and 5.7 times that on a 5° slope. Rainsplash becomes an increasingly important process, even though Morgan (1978) has shown that rainsplash alone accounts for less than 5 per cent of total erosion. Indeed, Kirkby and Kirkby (1974) record the splash displacement of a 5 mm particle of 150 mm but also observe for their Arizona sites that the seasonal growth of annual grasses reduces the efficacy of rainsplash. The processes of surface wash are of greater significance on shallow arid-zone slopes, as they are elsewhere (Williams, 1973).

In order to determine rates of soil loss and the retreat of these comparatively gentle slopes, troughs of a similar nature to those devised by Gerlach (1967) were installed on 4° and 15° straight slopes of well-defined length during the early rains of 1979. Three storms yielded 17 measures of wash erosion. There is a striking difference between the size distribution of the surficial soil of the erosion plots and the material washed off by overland flow (Figure 9.6). A typical median diameter for eroded soil is 0.2 mm which compares with 5 mm for the residual material. It is clear that material greater than 4 mm in diameter is unmoved, and that it is being concentrated in the regolith. This is the stone lag typical of shallow slopes in arid zones, and is similar to that reported by Yair and Klein (1973) from a study in the southern Negev. It will be shown later that this concentrating process is a significant control on the distribution of archaeological remains and vertebrate fossils.

Slope retreat

The rates of slope retreat calculated from the measurements in the Turkana Basin vary according to slope angle. They range from 0.2 mm a^{-1} for a 4° slope to 2 mm a^{-1} where a slope is 30° (Figure 9.7). This clearcut relationship contrasts with those reported for other arid zones where Kirkby and Kirkby (1974) for Arizona, and Yair and Klein (1973) for Israel indicate either no slope dependence or an inverse relationship attributed to either greater surface roughness and its influence on overland flow or the reduced mobility of coarser lag particles on steeper slopes. However, it conforms

Figure 9.6: Cumulative percentage size distributions of the residual surface soil of erosion plots on shallow straight slopes of 4-15° and the surface wash material carried to collecting troughs by overland flow. Triordinate graph illustrates the distinct concentration of coarser lag particles, in the residual soil, as fine particles are selected by the entrainment process.

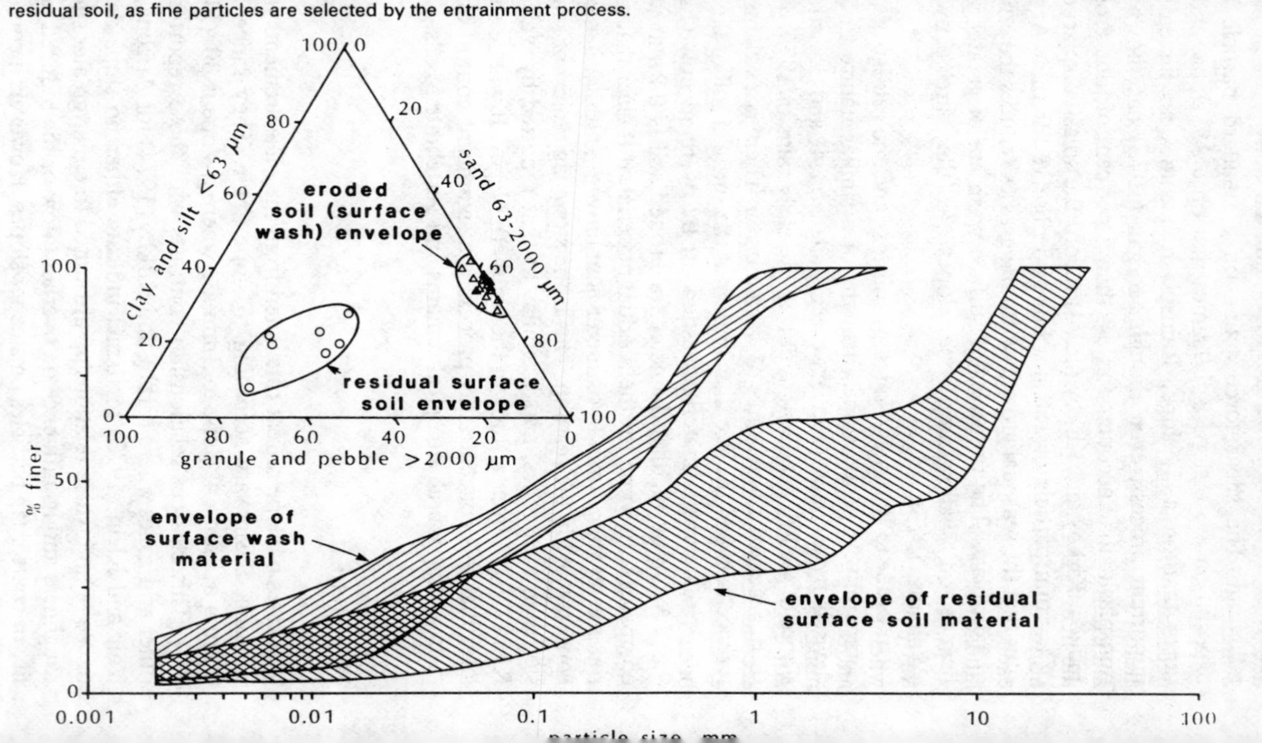

with the slope dependent erosion of sandy and silty clay soils determined from tree root exposure by Dunne *et al.* (1978) in Southern Kenya.

Annual rates of slope retreat in the Turkana Basin are not necessarily exceptional when viewed against those of other arid areas (Table 9.1). This is undoubtedly because the number of rainfall events per year is low due to the region's position peripheral to the major drought zones of the Ogaden and Sahel. A specific location might expect no more than six rainfalls in an entire year. But each storm is very effective in causing high rates of erosion and the landscape is therefore very dynamic.

Many carefully executed process studies do not necessarily throw light on the nature of slope retreat, nor do they establish whether slope evolution differs from one morphoclimatic zone to another. The present study attempts to bridge these gaps by examining the sideslopes of an expanding stream network (Figure 9.1). An evolutionary sequence of development is presumed in which second-order stream segments are older than the headward extending first-order fingertips, third-order stream segments have been developing longer than the second-order segments, and so

Figure 9.7: Slope-angle-dependent slope retreat displayed for shallow straight slopes with silty soils (circles) where rainsplash and surface wash dominate and for steep straight slopes of coarse talluvium (triangle) where gravitational processes are most active.

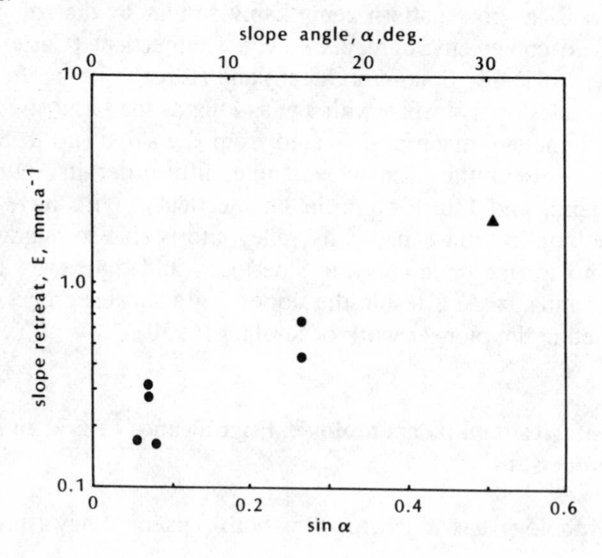

Table 9.1: Rates of slope retreat by environment

Reference	Environment	Slope (deg.)	Slope retreat (mm a^{-1})
Young (1960), *Nature, 188,* 120-2	Temperate Humid	Various	0.005-0.06
Kirkby (1967), *J. Geol., 75,* 359-78		13	0.0009
Williams (1973), *Austr. Geogr. Stud., 11,* 62-78		2-15	1-19
Koreleski (1974), *Prz. Geogr. 46,* 115-20		25	0-10
Williams (1973), *op. cit.*	Tropical Humid	0.5-14	5.6
Young (1977), *Catena, 4,* 289-307		Various	0.13
Young, (1960), *op. cit.*	Semi-arid	Various	2-8
Leopold *et al.*, (1966), *US Geol. Surf. Prof. Pap., 252G*		21-38	25
Yair and Gerson (1974) *Z. Geomorphol. Supp. 21,* 202-15		26-28	1-2
Campbell (1974) *Z. Geomorphol. Supp. 21,* 122-37		4-47	0-7
Barendregt and Ongley (1979) *Can. J. Earth Sci., 16,* 224-9		2-70	7
Frostick and Reid (1982), *Z. Geomorphol. Supp. 44,* 53-67		30	2

on. A plot of valley sideslope angle against stream order (Figure 9.8) reveals a gross pattern remarkably similar to that of much wetter and cooler environments — e.g. Connecticut (Carter and Chorley, 1961) and Southern Queensland (Arnett, 1971). In general, the sideslopes steepen with age as long as the basal stream is actively removing material delivered from the slope (up to third-order streams in this Kenyan example, fifth-order in Southern Queensland, and fourth-order in Connecticut). With increasing age, the basal stream expands its valley and is able to migrate so that it no longer undercuts the sideslopes and may even leave terrace remnants. As a result, the slopes degrade. This process was also noted in the pioneer work of Strahler (1950).

Archaeological and palaeontological significance of arid zone slope processes

Geomorphology has much to offer both palaeontology (through

Figure 9.8: Pattern of slope development in expanding valley systems from three distinctly different environments. (Data for S. Queensland from Arnett (1971), for Connecticut from Carter and Chorley (1961).)

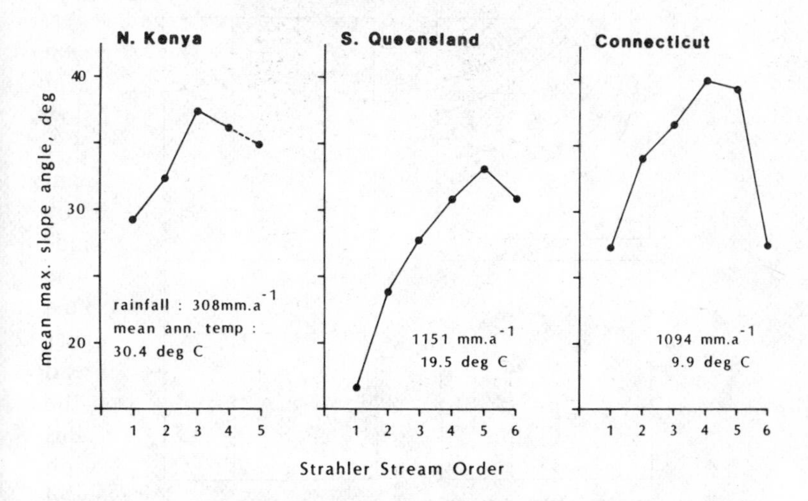

taphonomy) and archaeology as can be seen especially through the work of Butzer (1982). Changing slope form and slope processes have been used by Kirkby and Kirkby (1976) to assist in the dating of relatively recent sites of occupation in the Oaxaca Valley of Mexico and to explain the pattern of scatter of artefacts. Equally, work in Northern Kenya (Frostick and Reid, 1983) has not only elucidated natural processes but has also been used to support the palaeo-environmental reconstruction associated with the prolific early hominid finds of the area (Leakey, 1981). The aim of applying geomorphological techniques to archaeological and taphonomic problems in the Turkana Basin has been, in part, to determine the fate of fossils or artefacts exhumed by slope processes — the only way that new material comes to light — and to assess how far an object has slipped from its original stratigraphical position.

Fossilised bone and silicious artefacts (mainly core scrapers) were collected and marked. They were then treated as sedimentary particles and their long (a) intermediate (b) and short (c) axes ratioed to place them on a standard particle-shape diagram (Figure 9.9). The core scrapers are predictably either discs or blades. Choppers are more spherical. Of the fossils, limbs are, of course, exclusively rods, and vertebrae are spheres or discs. Scapulae are, however, not always blades.

Figure 9.9: Shape and sphericity, $[\sqrt[3]{(bc/a^2)}]$, of Plio-Pleistocene artefacts and fossils from the Koobi Fora Sedimentary Basin, N. Kenya.

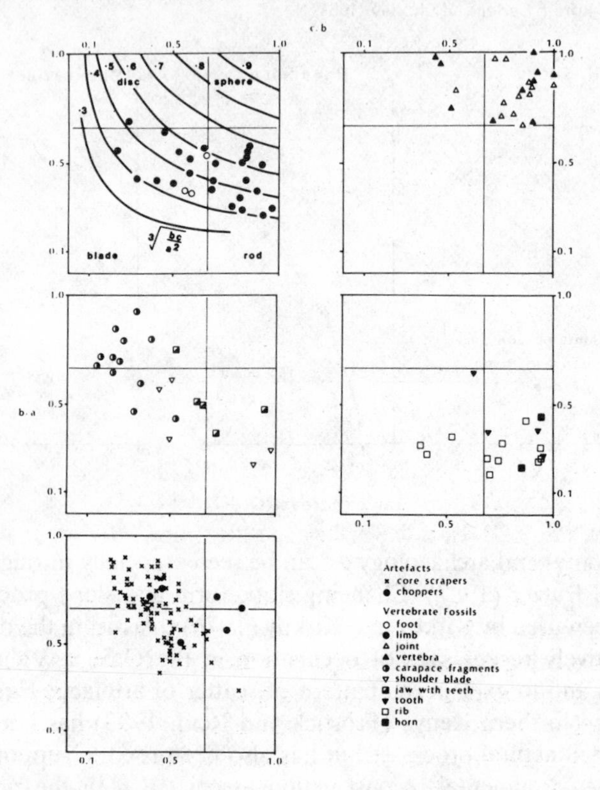

The material was carefully placed in appropriate attitudes within the talluvium on the 30° straight steep slope along with 2011 basalt particles, whose movements have already been described. Just as shape dictates a difference in recorded movement between the spherical Turkana particles and the platey Colorado clasts of Schumm (1967) so it is important in determining the downslope transport of the fossil bones and artefacts. Figure 9.10 indicates a clear difference in the sphericity of those fossils which move after two years of weather (Median value, ψ_{50} = 0.55) and those which remain stationary (ψ_{50} = 0.47). Rod-shape fossils undoubtedly move most quickly (averaging 1.85 m a^{-1}) followed by spheres (1.14 m a^{-1}), and discs (1.16 m a^{-1}) and then blades (0.57 m a^{-1}). Because of their shape, artefacts move at

Figure 9.10: Frequency distributions of sphericity (and hence rollability) of fossils seeded on the 30° talluvial slope that moved and remained stationary over two years of weather.

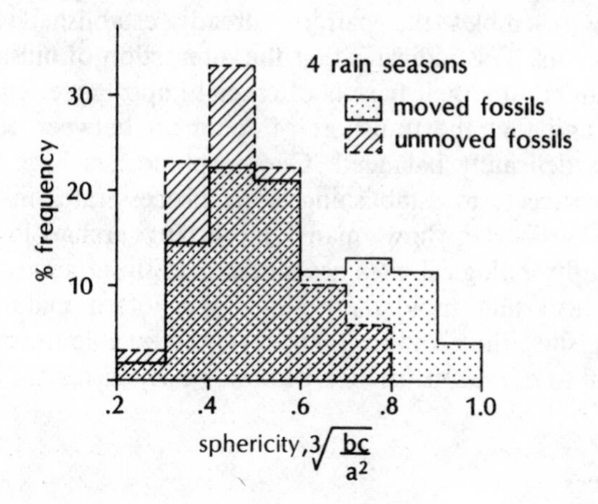

the slow rate of 1.13 m a^{-1} if they are discs and 0.07 m a^{-1} if they are blades.

Since the archaeological and palaeontological material is moving in general at half the rate of the rest of the talluvium it is obviously being concentrated. This is another reason why actively eroding slopes appear good collecting grounds. On gentler slopes, the tendency of coarse particles to remain as a lag gravel further accentuates the process of concentration. This has considerable implications for the practice of making palaeo-ecological reconstructions from fossil assemblages because material moving at different rates downslope from superimposed strata often may be the product of varying environments may become intermixed at the surface. Besides this, the fossil assemblage may become statistically biased as differently shaped elements escape from the slope mantle at varying rates of movement (Frostick and Reid, 1983) and become unrepresentative of the fossil fauna.

Conclusions

A wide range of erosion rates can be expected in any one type environment. However, differences can be explained by invoking fundamental mechanical principles that rely upon physical parameters such as particle mass and shape. What emerges from the Turkana Basin studies is that rates of slope retreat are directly related to slope angle. However, the nature of slope development with age resembles the pattern already established for other environments. This indicates that the interaction of hillslopes with the stream cutting their base is of crucial importance regardless of climate, and that the transport of sediment between slopes and stream is delicately balanced. Geomorphologists have had considerable success in establishing slope process-rates in semi-arid zones. Considering how many important archaeological and palaeo-anthropological sites are situated in these areas, and considering also that rapidly eroding slopes often make the best collecting sites, the geomorphologist should be able to contribute a great deal to the reconstruction of man's early history.

References

Arnett, R.R. (1971), 'Slope form and geomorphological process: an Australian example', *Spec. Publ. Inst. Brit. Geogr.*, *3*, 81-92

Bakker, J.P. and Le Heux, J.W.N. (1952), 'A remarkable new geomorphological law', *Koninklijke Nederlandsche Akad. van Wetenschappen Series B*, *53*, 1073-84; 1364-74.

Butzer, K.W. (1982), *Environment and Archaeology* (Methuen, London), 524pp.

Caine, N. (1981), 'A source of bias in rates of surface soil movement as estimated from marked particles', *Earth Surf. Proc. and Landf.*, 6, 69-75.

Campbell, I.A. (1974), 'Measurements of erosion on badlands surfaces', *Z. Geomorph. Suppl.*, *21*, 122-37

Carter, C.S. and Chorley, R.J. (1961), 'Early slope development in an expanding stream system', *Geol. Mag.*, *98*, 116-30.

Davis, W.M. (1932), 'Piedmont bench lands and Primaerruempfe', *Geol. Soc. Amer. Bull.*, *42*, 399-440.

Dunne, T., Dietrich, W.E. and Brunengo, M.J. (1978), 'Recent and past erosion rates in semi-arid Kenya', *Z. Geomorphol. Suppl.*, *29*, 130-40

Dury, G.H. (1959), *The Face of the Earth* (Penguin, Harmondsworth), 255 pp.

Frostick, L.E. and Reid, I. (1982), 'Talluvial processes, mass wasting and slope evolution in arid environments', *Z. Geomorphol. Suppl.*, *44*, 53-67.

Frostick, L.E. and Reid, I. (1983), 'Taphonomic significance of sub-aerial transport of vertebrate fossils in steep semi-arid slopes', *Lethaia*, *16*, 157-64

Frostick, L.E., Reid, I. and Layman, J.T. (1983), 'Changing size distribution of suspended sediment in arid-zone flash floods', *Spec. Publ. Int. Assoc. Sediment.*, *6*, 97-106.

Gerlach, T. (1967), 'Hillslope troughs for measuring sediment movement', *Rév. Geomorphol. Dyna.*, *17*, 173.

King, L.C. (1967), *The Morphology of the Earth* (Hafner, New York), 726 pp.

Kirkby, M.J. (1967) 'Measurement and theory of soil creep', *J. Geol.*, *75*, 359-378.

Kirkby, A. and Kirkby, M.J. (1974) 'Surface wash at the semi-arid break in slope', *Z. Geomorphol. Suppl.*, *21*, 151-76.

Kirkby, A. and Kirkby, M.J. (1976), 'Geomorphic processes and the surface survey of archaeological sites in semi-arid areas', in Davidson, D.A. and Shackley, M.L. (eds), *Geoarchaeology* (Duckworth, London), pp. 229-53.

Leakey, R.E. (1981), *The Making of Mankind* (Michael Joseph, London), 256 pp.

Leopold, L.B., Emmett, W.W. and Myrick, R.M. (1966) 'Channel and hillslope processes in a semiarid area, New Mexico', *US Geol. Surv. Prof. Pap., 352G.*

Melton, M.A. (1960), 'Intravalley variation in slope angles related to micro-climate and erosional environment', *Geol. Soc. Am. Bull., 71*, 133-44.

Morgan, R.P.C. (1978) 'Field studies of rainsplash erosion', *Earth Surf. Proc., 3*, 295-9.

Penck, W. (1924) *Die morphologische Analyse: ein Kapitel der physikalischen Geologie* (Stuttgart) trans. Czech, H. and Boswell, K.C. (1953), *Morphological Analysis of Landforms* (Macmillan, London).

Ruxton, B.P. (1958) 'Weathering and subsurface erosion in granite at the piedmont angle, Balos, Sudan', *Geol. Mag., 95*, 353-77.

Saunders, I. and Young, A. (1983), 'Rates of surface processes on slopes, slope retreat, and denudation', *Earth Surf. Proc. and Landf., 8*, 473-501.

Savigear, R.A.G. (1952) 'Some observations on slope development in South Wales', *Trans. Inst. Brit. Geogr., 18*, 31-52.

Schumm, S.A. (1956a), 'Evolution of drainage systems and slopes in badlands at Perth Amboy, N. Jersey', *Bull. Geol. Soc. Am., 67*, 597-646.

Schumm, S.A. (1956b), 'The role of creep and rainwash on the retreat of badland slopes', *Am. J. Sci., 254*, 693-706.

Schumm, S.A. (1964), 'Seasonal variations of erosion rates and processes on hillslopes in W. Colorado', *Z. Geomorphol. Suppl., 5*, 215-38.

Schumm, S.A. (1967), 'Rates of surficial rock creep on hillslopes in Western Colorado', *Science, 155*, 560-1.

Strahler, A.N. (1950), 'Equilibrium theory of erosional slopes approached by frequency distribution analysis', *Am. J. Sci.*, 248, 673-96.

Williams, M.A.J. (1973), 'The efficacy of creep and slope wash in tropical and temperate Australia', *Austral. Geogr., Stud., 11*, 62-78.

Wood, A. (1942) 'The development of hillside slopes', *Proc. Geol. Assoc., 53*, 128-40.

Yair, A. and Klein, M. (1973), 'The influence of surface properties on flow and erosion processes on debris-covered slopes in an arid area', *Catena*, 1, 1-18.

Yair, A. and Lavee, H. (1976), 'Runoff generative process and runoff yield from arid talus mantelled slopes', *Earth Surf. Proc., 1*, 235-47.

Yair, A., Lavee, H. Bryan, R.R. and Adar, F. (1980), 'Runoff and erosion processes and rates in the Zin Valley badlands, N. Negev, Israel', *Earth Surf. Proc., 5*, 205-225.

10 KARST FORMS AND PROCESSES

A.F. Pitty, R.A. Halliwell and Helen S. Goldie

Introduction

The caves of karst areas have proved to be rich lodes of archaeological remains, due to the shelter and equable climate afforded to wild animals and to early man. Caves occupy a special niche in some modern as well as more primitive religions, folklore and superstitions. At the beginning of the seventeenth century, palaeotourists were already trekking to the Vilenica cave in Slovenia to experience an awe of the underworld (Gams, 1974, p. 179). Today, major caves in post-industrialised societies annually attract visitors by the million, and thousands of speleologists explore their inner reaches. In consequence, the theme which karst geomorphology is perhaps most conspicuously suited to demonstrate is the involvement of the general public with landforms. This chapter, therefore, focuses on how karstforms can satisfy the urge to explore the unknown for a wide range of natural abilities, why the aesthetic values of karstforms are worth conserving, and on the explanations for their curious shapes. For many on school field trips, in the North of England at least, karstforms are amongst the first to arouse a scientific interest in landforms.

Cave exploration

First short crawls

When early hominids' technology had advanced enough to provide burning resinous branches, and later stone lamps burning oil or fat, they could probe further into caves to where their famous cave paintings now remain. Exploration with equipment in the form of boats, ropes and rigid ladders began in the sixteenth century. The use of hand winches in the eighteenth century is recorded, with rope ladders being used in France in 1780. In Yorkshire, the earliest descriptions of caves were probably those of R. Pococke (1751) and J. Hutton (1781) who wrote in his *A Tour to the Caves* that 'what is most extraordinary, these subterranean brooks cross each other underground without mixing waters, the bed of one being on a stratum above the other: this was discovered by the muddy water after a sheep washing, going down the one passage, and the seeds or husks of oats that were sent down the other' (Halliwell, 1979).

Nineteenth-century expansions underground

From the 1830s onwards, cave exploration increased on a large scale, following the Rev. J. Buckland's sensational discovery of ancient animal remains in East Yorkshire. Exploration in North-west Yorkshire was led by local landowners, including J. Birkbeck. In 1847, he led a group of ten to Alum Pot who 'ventured into this awful chasm with no other apparatus than ropes, planks, a turn-tree, and a fire-escape belt'. Pioneer geologist, J. Phillips (1855) published measurements of the growth of stalagmites in Ingle-borough Cave, a list enlarged on by J. Boyd Dawkins (1874) in *Cave Hunting*, the first book in English devoted entirely to cave science. In the 1890s, the Yorkshire Ramblers Club, although its constitution did not mention caving, became a major driving force in cave exploration. This is a significant example of how interests in landforms are channelled by the particular landscapes addressed. The club's members included Cuttriss, who, on reaching the bottom of a pot 'always accompanied with that mysterious green rucksack of his — busied himself with chipping off bits of rock and taking the temperature of the air and water, but I could not for the life of me see why in doing this, he should consider it necessary to stand up to his knees in water for ten minutes or so, with a thermometer dangling by a bit of string from the button of his coat. The compass and barometer, too, has to be consulted.' However, the club's aspirations to descend Gaping Gill, the deepest vertical shaft in the UK, like R.F. Scott's later trek to the South Pole, were anticipated by a fleet-of-foot foreigner. In 1895, a co-founder of karst geomorphology, E.A. Martel, took only 23 minutes to climb some 100 metres down his rope ladder to the floor of the Main Chamber (Figure 10.1).

Twentieth-century technological advances

By the 1930s, some of the most severely testing cave passages were opened up, notably Swinsto Hole. Eli Simpson generalised on his experiences, including some 17 years crawling up and down Swin-sto, in an important paper, 'Notes on the formation of Yorkshire

Figure 10.1: Martel about to descend Gaping Gill.

caves and potholes'. This stressed the importance of shale beds in cave development in North-west Yorkshire and offered an explanation for the long wet crawls which exploration was encountering increasingly. In the austerity years of the 1950s there was another boom in inexpensive outdoor pursuits. In addition, cavers had the challenging thesis (Sweeting, 1950) that caves tended to occur at specific levels, related to former baselevels, on which to focus their thoughts. Exploration was rapidly pushed forward, both with advances into more difficult caves, such as Penyghent, and with intensive digging to find new caves. In the 1960s, exploration was boosted by the introduction of wet suits and lightweight metal ladders. As these replaced bulky clothes and even bulkier rope ladders, much smaller, previously 'impossible' cave passages could be explored. Bulk has been reduced still further by the adoption of rockclimbers' single-rope techniques (SRT). This was also the time of the emergence — and submergence — of the cave diver. Earlier diving explorations had involved bulky breathing apparatus but, by the early 1970s, the first dive of over 300 m had been achieved using lightweight scuba gear. The use of diving gear has been greatly extended during the last decade and recent feats include diving the 1830 m connection from Kingsdale Master Cave to Keld Head. In 1983, the 90-year awaited Gaping Gill–Ingleborough Cave link up was finally achieved. This link-up depended on the use of the 'Molephone', a radio frequency inductive-loop system which facilitates cableless, two-way voice communication between ground surface and cave. Also, thanks to the skills of a TV cameraman and the teams of 'speleosherpas' carrying the specially designed lighting equipment, millions have participated in such unique events from their armchairs.

Conservation of karstforms

Concern for features of geomorphological and geological interest has increased in recent decades because of greater pressure on our environment. Conservation efforts have therefore had to increase, though often they encounter society's unwillingness or inability impartially to assess the value of these features in relation to more quantifiable economic uses of the materials or land involved. A specific karst example, the case of limestone pavements, will illustrate some facets of the landform conservation issue.

Description and uses of limestone pavements

Limestone pavements are exposed, glacially stripped surfaces of hard limestone which are composed of blocks or *clints* of rock, separated by solution-widened crevices or *grikes* (Williams, 1966). Natural solution of such exposed limestone surfaces creates beautiful patterns of runnels on the clints (Figure 10.2). These patterns have attracted horticulturalists to remove and use the clints for garden rockeries. In recent years damage caused to the natural outcrops by this removal has been of great concern in areas valued for their natural beauty, notably in Britain and Ireland (Williams, 1966; Sweeting, in Herak and Stringfield, 1972; Goldie, 1973). Limestone pavements in the British Isles have also been greatly altered by a variety of economic activities over longer periods of time. Indirect effects include soil depletion due to grazing and cultivation. More directly, blocks have been removed for building walls and houses, surfacing roads, lime-burning, and for pasture improvement. All these alterations constitute a significant geomor-

Figure 10.2: Runnelled limestone pavement in the Great Ashby Scar area, Orton, Cumbria.

Figure 10.3: Damaged limestone pavement on Hampsfield Fell, Cumbria. The surface is roughened, lacks runnels, and is covered by loose, flakey clasts. The larger broken blocks in the background are spoil from the extraction process.

phological process, relevant to the scientific understanding of these microforms and of the distinct ecological niches which they contain. The botanical interest of limestone pavement lies with the distinctive shade-loving flora of the grikes. The possibility of legal protection for these features now exists in the United Kingdom, and has been urgently needed because of the accelerating rate of damage.

Methods of study

The sources and methods which identify such effects on pavements are varied (Goldie, 1976). In the field, areas of clint removal can be identified because the lowering of the limestone surface can increase the size of the blocks composing the remaining roughened surface and decrease the width and depth of the joint-controlled grikes which separate the clints (Figure 10.3). In addition, much loose gravel may be left after clint removal. Thus, the damaged pavement appears quite different from its natural condition and the character of the ecological niche is drastically changed. Small damaged areas can be found at virtually all pavement outcrops in the British Isles. Yorkshire examples include Scales Moor on Whernside, where damaged pavement has been grassed over, and Gauber Pasture on Ingleborough, where damage is old and the pavement is partially grassed over. Examples in Cumbria include Hampsfield Fell, where runnelled clints have been heaped up by the contractors; Gaitbarrows, where large areas of thick beds of limestone have been removed; and Gaythorne Plain, once a pavement but now gravel and broken rock on a rough limestone surface, with virtually no identifiable smooth or runnelled clints. In Ireland, around Dun Dhubcathaire stone fort on Arainn, a mixture of lowered, roughened clints and higher, smooth clints, indicates where removal has occurred. In addition to basic field evidence, historical sources, such as Inclosure Awards, the records of local estates, old newspapers, aerial photographs and archaeological literature can all be useful. More recently, records and files of the Nature Conservancy Council include botanical survey material for the British pavements, and much local correspondence on damaged sites.

Depositional patterns

As with any denudational process, evidence of the scales, rates and duration of clast removal can be deduced from depositional patterns. Where human activity is the denudational agent the length of its duration is often striking. Limestone pavement blocks can be identified in structures as old as the fourth century AD Romano British village at Din Lligwy on Anglesey and prehistoric hut walls, such as at Oxenber Wood in Yorkshire, and in stone circles, for example at Knipe Scar in Cumbria. At Din Lligwy, the huts of the village are actually built on *in situ* pavement, using clints which have been moved to make the walls and gateposts. In Ireland, burial chambers such as Poulnabrone Dolmen in the Burren, Co. Clare, have clearly used clints in their construction. Several large stone forts on the near pavement outcrops on the island of Arann, off Co. Galway, have been made at least partly of pavement blocks. Clints have also been used in defensive lines, or 'chevaux de frises', around these forts, for example, Dun Aonghasa.

Solution-sculpted blocks from pavements decorate the tops of walls in villages and towns, such as Kendal, Ingleton and Long Preston in the North of England. This usage may be systematic along whole streets, and some larger clints have been used as gateposts. In one early twentieth-century Kendal street, every house has two clint gateposts as well as decorative pieces.

Pressures on pavements

Today the main demand for pavement blocks is for constructing garden rockeries. Horticulturalists first used pavement blocks, taken from the Ingleborough area, more than 100 years ago. In addition to their aesthetically pleasing shapes, the clints are valued because artificial streams can follow courses down the runnels (Minney, 1983). This elegant art is seen at Sizergh Castle Cumbria, where the rock garden, built in the early twentieth century occupies an extensive, flattish area through which a stream flows. The clints have been used as bridges, provide channels for the waterflow, and are used as ledges in low, stepped rockeries. The runnels however, become masked by rockery vegetation, a factor not realised in plans for runnelled clints in gardens; another unappreciated fact is that modern extraction methods using explosives, hydraulic shovels and other heavy equipment fracture clints, so gardeners rarely get the features they desired. As clints are extracted not only is there damage to the clints being removed, but also to the remaining ground. This contrasts with the methods used at the turn of the century when clints were extracted carefully and selectively by hand and crowbar. This simple change typifies how landform modification by man has taken a technological leap in recent decades, and why there needs to be a swifter and stronger action by conservationists to prevent further damage. Features in magazines and usage in public places stimulate demand, however, as has been the case since the exhibitions at Crystal Palace in 1863 and 1911. The Festival of Britain in 1951, and the 1981 Chelsea Flower Show also featured artificial rockeries made of natural clints. Landscape rockeries using clints can also be seen in other public places, such as West Midlands factory entrances, museum gardens, a traffic island in York, and in London squares and hotel gardens.

Larger-scale landscaping may also transform natural limestone rock surfaces. For example, in Yugoslavia, on the Kanin plateau above Bovec in the Julian Alps, a valley side of scree has been bulldozed to make ski runs on the valley floor. As a result, pavements have been covered and dolines filled in with gravel taken from the screes. Possibly the largest scale obliterations of pavement, however, are the result of unrestricted quarrying of limestone for roadstone and the cement industry, as in the Peak District of Derbyshire and on Mendip (Stanton, 1982) where caves have been lost, or damaged. Similarly, pavements in the Yorkshire Dales and Cumbria have been quarried away.

Reducing the pressures

Clearly, a worthwhile geomorphological task is to highlight the risks of losing scientifically interesting, ecologically critical and aesthetically pleasing landforms and microfeatures. There seem to be four effective strategies for reducing pressure on these features. First, if introduced in time, voluntary control can be effective simply by making the exploiter aware of the damage being done. For instance, at Holme quarry, near Milnthorpe, Cumbria, an area of pavement with unique features has been lost, but the remainder of good quality is now protected. Secondly, there is legislation for protection to be compulsory if landowners do not comply voluntarily, and has been implemented at Hampsfield Fell, Cumbria, in the first Limestone Pavement Order under the Wildlife and Countryside Act, 1981. Thirdly, protection can be achieved if valuable sites are acquired and conserved as designated nature reserves by bodies such as the Nature Conservancy Council and local nature conservation trusts. Examples are Gaitbarrows, Cumbria, and Scar Close, Yorkshire. Fourthly, the public should be encouraged to view pavements in their natural surroundings, rather than to destroy them to make artificial rockeries. The damage done by removal, the destruction of habitats for rare plants, the removal of unique landforms which are also beautiful landscape features but which cannot be recreated in gardens can all be demonstrated to the public both in the field and with photographic evidence. There is little pavement left in Britain which is undamaged, and about half the total 2150 hectares has already been destroyed (Ward and Evans, 1976). The public needs to be shown that these features are rare and irreplacable, like stalactites in caves (Black, 1969).

Enclosed depressions in karst

With caves hidden from view, and pavements comparatively rare, it is basin- or funnel-shaped enclosed depressions which are the most characteristic landform of karstlands. In fact, Prestwich (1854, p. 222) wrote 'The occurrence of swallow holes on chalk and limestone hills is a phaenomenon almost too well known and too general to call for any special notice.' In the eighteenth century, these distinctive forms were attributed to the collapse of caves, in keeping with the Catastrophists' theme of earth history. However, by the mid-nineteenth century, observations by Prestwich in chalklands of southern England, and by David Dale Owen in Kentucky, supported Charles Lyell's view that enclosed depressions were formed gradually by long-continued solution by infiltrating water (Roglic, in Herak and Stringfield, 1972, p. 3). This view of earth history change is termed *gradualism.*

Much karst research since the mid-twentieth century has taken up the theme of *actualism,* by calculating the approximate rates by which present-day processes operate and using these to indicate orders of change over longer periods. Generally, 70 per cent or more of this denudation takes place at subsoil surfaces of limestone (Ford and Drake, 1982, p. 149). The reasons why certain karstforms develop their distinctive shapes and patterns are less generally agreed. The following discussion focuses on how geological circumstances direct water movement at particular points or zones, or may have done so in the past, and thus exemplifies the theme of geological control on landform development (Pitty, 1984).

Enclosed depressions in North-west Yorkshire

In North-west Yorkshire, the most extensive karst area in the British Isles, sinkholes are common along, or close to, the margin of the shales which overlie the limstone. As in many other karst areas, this pattern is obviously related to the headwaters which have gathered on impermeable rocks, disappearing underground after they have flowed onto the pervious limestone. This situation, however, is difficult to envisage for the much larger and less regularly shaped enclosed depressions, concentrated on top of the Parson's Pulpit–High Mark plateau, north of Malham. They are remote from areas of higher land and therefore unconnected with the present-day drainage from impermeable headwaters. There are 17 depressions on Parson's Pulpit and 12 on High Mark. Measuring up to 750 m in length, they are as large as any enclosed depressions in the UK (Moisley, 1955). Figure 10.4A is a representative slope profile on the north-west flank of a High Mark depression. Apart from a slightly curved summit convexity, upslope from the minor step at the outcrop which rims the depression, the slope gradient is conspicuously constant, and the corresponding line graph shows that the downslope sequence of angles is distributed equally about the mean of 9° (Figure 10.4B). In the corresponding symmetrical bar graph, 51 per cent of all observations fall in the 6-11.5° range, with 18.6 per cent concentrated in the 10-11° modal class. However, the scatter of angles increases downslope, indicating an increase in ground surface roughness which is even perceptible in the profile. Calculating an index from the four successive differences within consecutive 5-angle groups suggests that this progressive trend is highly significant. The field observations noted in Figure 10.4B record two prominent characteristics of the slope: (i) bioturbation is intense in such sheltered hollows below the exposed plateau top; and (ii) tunnelled spoil is readily displaced downslope by gravity and rainbeat. About 70 m downslope, spoil is sufficiently thick and widely distributed to smooth the slope surface. Further downslope, however, limestone blocks increasingly protrude above the general level of the interspersed patches of soil. Thus, despite the activity of downslope movement of debris and examples noted elsewhere of debris accumulating in karst basin floors (Crowther and Pitty, 1983, p. 72), debris is apparently evacuated from the base of the High Mark depressions. This evacuation is suggested, equally, by the arenaceous sediments in the caves on the northern flank of the upland massif (Long in Waltham, 1970). Similarly, fine, bleached sand is found at risings around the perimeter of the upland. It seems that depressions, once formed, can be gradually enlarged by present-day processes. However, since a limestone block near the centre of the depression has solutional features developed in accordance with its present attitude, eluviation and any basin floor settling is clearly a very slow process.

Whereas flattening of the basin flanks may be the predominant present-day process, these depressions were probably initiated by solution, as water percolated through a former, overlying sandstone cover (O'Connor *et al.*, in Waltham, 1970, p. 399). The anomalous gritstone block, noted near the slope profile could be a remnant from the former cover, lowered from its originally higher stratigraphic position. Even if it is a glacial 'erratic', the widespread occurrence of sandstone fragments in the soil equally suggests the former presence of a permeable cover above the limestone at this locality. The geological history of this stratigraphical succession could also explain the geographical location of these enclosed depressions. This succession is developed within a narrow zone on the flanks of Wharfedale, to the east and west of Grassington, due to the top of the Great Scar Limestone being an *unconformity*; that is, a geologically ancient surface, at a pause in deposition, possibly with erosion too. Earth movements were commonly involved, and strata above an unconformity usually make an angle with the attitude of the older rocks. In the present case, Carboniferous sedimentation, which occurred as the limestone block gradually submerged, was due to periodic incursions of deltas from the North. However, the southern part of the block was land, or at least an area of non-deposition, until sandstones of the Grassington Grit were deposited (Rowell and Scanlon, 1958, p. 87). Therefore, the limestone in a narrow East-West zone, just north of the Craven Faults, was succeeded directly by sandstone, either of the Grassington Grit group or, a little to the North, by the sandstones of the underlying Yoredale series (Figure 10.5A). Elsewhere in Northwest Yorkshire, the Great Scar limestone is usually succeeded by shales.

Figure 10.4: Slope-profile analysis of the flank of an enclosed depression on High Mark, north of Malham. A. The slope profile (natural scale). B. Line-and-bar graph of the sequence of angles in the profile. C. The downslope trend in ground surface roughness.

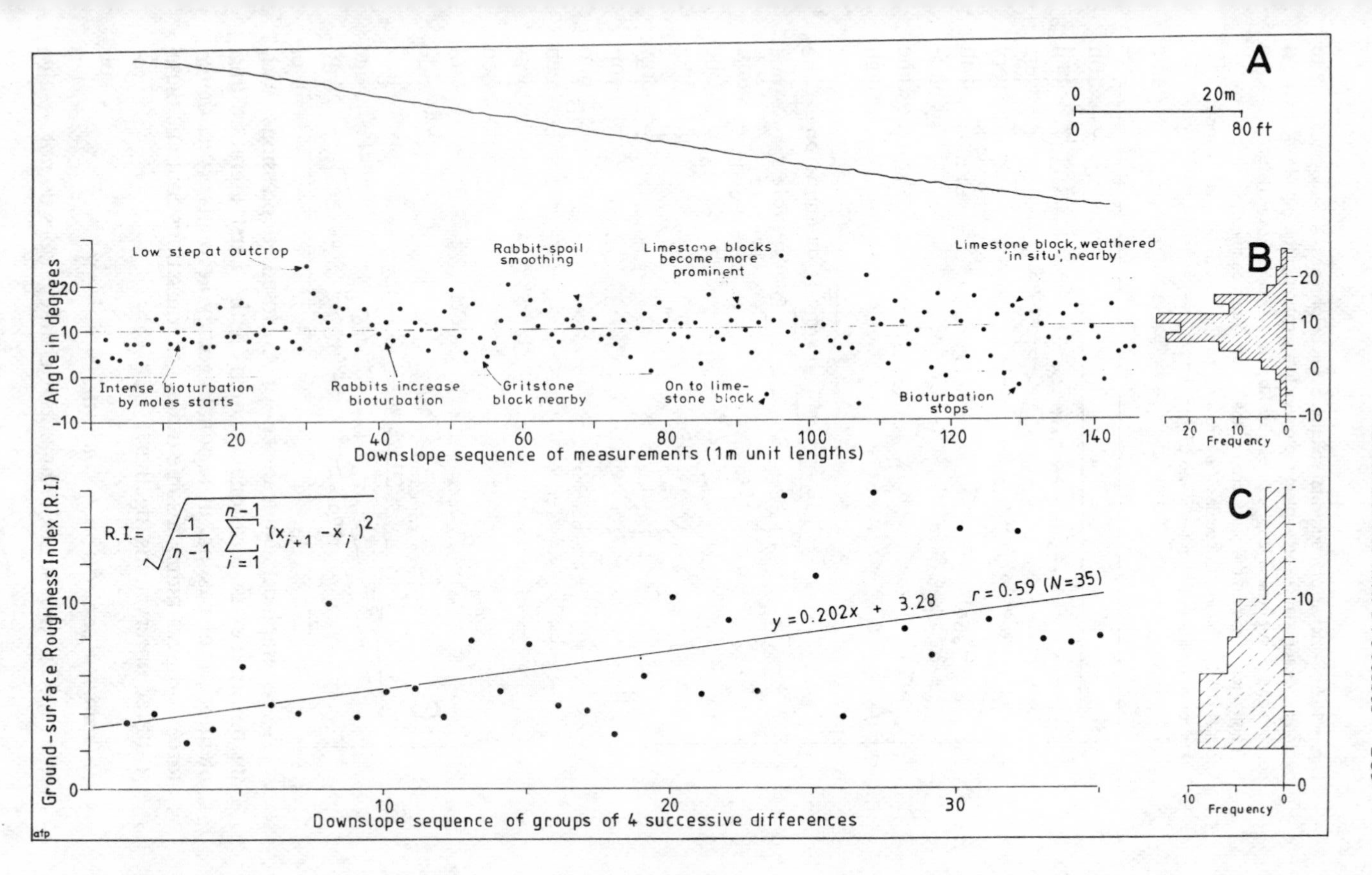
A
0
20m
0
80 ft
Low step at outcrop
Rabbit-spoil smoothing
Limestone blocks become more prominent
Limestone block, weathered 'in situ', nearby
B
Angle in degrees
Intense bioturbation by moles starts
Rabbits increase bioturbation
Gritstone block nearby
On to lime-stone block
Bioturbation stops
Downslope sequence of measurements (1m unit lengths)
Frequency
Ground-surface Roughness Index (R.I.)
R.I. = $\sqrt{\frac{1}{n-1}\sum_{i=1}^{n-1}(x_{i+1}-x_i)^2}$
C
y = 0.202x + 3.28
r = 0.59 (N=35)
Downslope sequence of groups of 4 successive differences
Frequency
atp

Figure 10.5: Sketches of interstratal solution at unconformities. A. Suggested geological circumstances to explain the distinctive location of the High Mark depressions. B. A Furness 'sop', as described by Dunham and Rose (1949). C. Solution subsidence on the north crop of the South Wales coalfield, studied by Thomas (1963). The vertical scales are notional and the horizontal scale is about 10 km for A, and about 1 km for B and C.

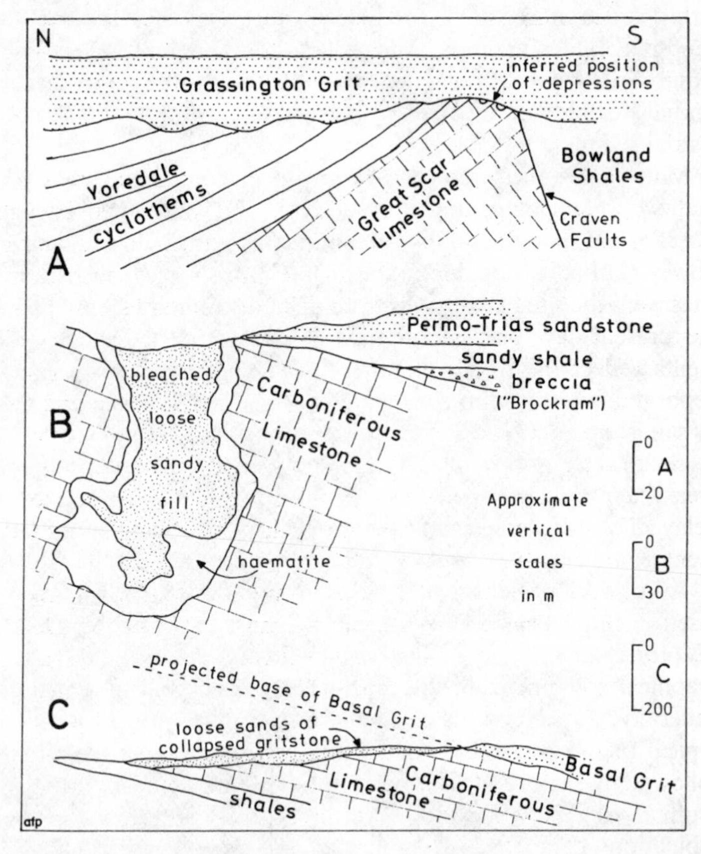

A geographical coincidence of enclosed depressions with unconformities is occasionally noted in the karst literature and mentioned, in passing, in regional geology accounts. It is, therefore, worth examining if some useful generalisations could tie these scattered observations together.

Lithological controls on groundwater movement

In a succession of strata, sedimentary characteristics do not usually change at random. With continuous sedimentation, there is a progressive development or systematic change in lithology. Limestones are usually overlain by shales or mudstones, partly due to the distinctive environment required for limestone accumulation. Since debris-free and calm water is needed for precipitation of calcium carbonate, it follows that any gradual change will involve turbid conditions, with mud-sized particles the first to be deposited. Furthermore, since limestone deposition usually requires a warm environment, high evapotranspiration on adjacent landmasses reduces the ability of streams to evacuate coarser terrestrial sediment to the sea.

Many limestones were laid down in 'epicontinental' seas (the shallow waters of continental shelves), and were then brought closer to old shorelines if the epicontinental seas shallowed. During Lower Carboniferous deposition in the British Isles, such shorelines were the site of overspills from delta distributaries, resembling the present-day 'bird's foot' type (Moore, 1959). Deltaic sediments reflect the depth of water in which they were originally deposited, and thus limestones are overlain first by shale and then by sandstone. Periodically, limestone deposition was re-established in clear water, following periodic subsidence in the depositional area when the Yoredale series were deposited, perhaps related to delayed isostatic adjustments to the weight of accumulating sediment (Dunham, 1950, p. 58).

An upward passage from limestones into shales is widely recognised in generalised stratigraphical columns. For instance, Duff *et al.* (1967) include 13 Pennsylvanian limestones in the 11 stratigraphical columns from the Appalachian Basin, all of which are generally followed by shale or underclay. Similarly, in 11 locations typical of the Scottish Carboniferous, there are seven limestones followed by shale, and a further two followed by mudstones. In all, Duff *et al.* (1967) include some three dozen limestones in their tabulations, each of which is followed by a shale or other mudrock.

The general tendency for mudrocks to be superimposed on limestones and, equally, the exceptions to this general rule, are cardinal features of karst geomorphology in structurally stable areas. Translating the stratigraphy into geohydrological terms, the water-bearing, pervious limestones are *aquifers*; and overlying shales, which can transmit water only at negligible rates, are *aquicludes.* Where exposed at the ground surface, the edge of the outcrop of such aquicludes is generally the zone where sinkholes are developed (Clayton, 1981), with gravity flow streams leading into the incised slots of *vadose* caves. If still buried at some depth, such aquicludes act as a *confining stratum*, in which water in the underlying limestone aquifer is contained under pressure, and tube-like *phreatic* caves are developed. In localities where the 'seal' of a confining aquiclude is first punctured by sub-aerial erosion, the groundwater rising to the surface under pressure may be warm and mineralised *spa water*. This dual role of aquicludes, both in preventing downward water movement from above and confining upward movement from below, is recognised, for example, in the Mendip karst area. Here, marl bands may throw out springs from overlying limestone conglomerates. Equally, in other structure-relief situations, these aquicludes may keep water down in the underlying, more pervious beds (Green and Welch, 1965, pp. 174-5).

Unconformities and foci of interstratal solution by gravity flow groundwater

The general rule that a mudrock tends to succeed a limestone, in either a shallowing or deepening sea, depends on sedimentation being continuous. However, locally different vertical successions in lithology can result at unconformities. These may be a significant geohydrological control if the mudrock which usually succeeds a limestone is missing and a sandstone or some other permeable formation rests directly on a limestone at the unconformity. With the aquiclude above the limestone absent, groundwater circulating in the aquifer above the limestone begins to dissolve limestone where the seal is absent. This process, termed *interstratal solution*, is particularly effective in concentrating solutional activity, simply due to the volume of flow which may converge on the leak in an aquiclude seal and to the length of time over which such convergences of groundwater flows may persist.

In the British Isles, the first instance described was that of the solutional widening of joints in chalk seen in many quarry faces, and attributed to long-continued contact with groundwater circulating in the overlying, slightly unconformable Eocene sands. The most decisive evidence about interstratal solution at an unconformity comes from the Low Furness area of north Lancashire (Figure 10.5B). North of Barrow-in-Furness, near the present northern margin of the Permo-Trias outcrop, overstepping St Bees sandstones rest directly on Carboniferous formations. Here, 'sops' are developed, deep natural shafts in the limestone and lined with high-quality iron-ore, deposited by water circulating from the sandstone into the limestone (Dunham and Rose, 1949). The deposition of the iron in the sops reflects the degree to which the flow of very ferruginous water was concentrated in relatively few channels. The elimination of apatite in the sandy cores of the sops, a common heavy mineral in the St Bees sandstone parent material, is another indicator of the concentration of water flow causing intensive solution over a protracted period of time. Conversely, no sops have been found where the St Bees shale aquiclude intervenes above the limestone.

Comparable haematite iron ores were deposited in the Carboniferous limestone of the Vale of Clwyd by downward percolating waters. Elsewhere in North Wales, due to contemporaneous erosion, Millstone grit overlies limestone unconformably and, about 400 m from its present-day outcrop, there are many natural shafts with large subsided blocks of Millstone grit (Walsh and Brown, 1971, p. 311). Similar solution-subsidence features are extensively developed in South Wales along the outcrop of the Millstone grit where it directly overlies the Carboniferous limestone. Here, with sub-surface solutional lowering of the limestone, the outer edge of the overlying Basal grit has subsided by up to 250 m, compared with its projected former elevation (Figure 10.5C). Thomas (1963) suggested that large caverns or 'halls' were developed by 'undersapping' beneath a water-table. He attributed the marked development of the related enclosed depressions (some 50-80 per hectare) to the particularly high rainfall in the hills of South-Central Wales and its collecting on the long dip-slopes of non-calcareous strata extending northwards from beneath the limestone horizons. The enclosed depressions were attributed to catastrophists' style collapses of the 'halls'.

Alternatively, a gradualist's view of solution and subsidence occurring concurrently over long periods of time can be envisaged. Since there is no angular unconformity at the marked stratigraphic gap between the limestone and succeeding gritstones, the area can be regarded as unusual in that an aquifer rests directly on top of the limestone over an extensive area.

The Mendip area, in contrast, is closer to the more general situation of localised punctures in the overlying aquiclude seals. For instance, in the Shepton Mallet–Wells area, groundwater passes directly into sub-surface Carboniferous limestone only where

Rhaetic shales or other impermeable strata do not intervene (Green and Welch, 1965, p. 175). In Western Ireland limestone is the only country rock over an extensive part of the North and East of the Co. Clare karst terrain. In consequence, there is no direct evidence on the nature of the former cover rocks. To the South, however, the Clare shales overlying the limestone thinned dramatically towards the northern edge of their present-day outcrop, from over 300 m to less than 12 m within a 50 km distance. Clearly, the northern limit of this depositional zone could not have extended much further north (Hodson and Lewarne, 1961, p. 314). Later, submarine alluvial fans advanced and deltaic sandstones covered the area, including the local trenching of the sub-sandstone shale by channels up to 20 m deep. Conceivably, therefore, sandstones might have rested directly on the limestone in the central and eastern areas where the enclosed depressions are developed. Indeed, there are no enclosed depressions in the southern and south-western parts of the Clare limestone terrain (Sweeting, 1955, p. 47), where outliers of shale are present and a formerly continuous aquiclude is, therefore, readily envisaged.

Some other aspects of enclosed depressions

A key to understanding enclosed depressions is 'convergence', that is, the geomorphological concept that contrasted denudational mechanisms can eventually fashion similar landforms. Thus, in addition to the generalisations and cases outlined so far, such karst cavities which have developed beneath a shale horizon in the Joliet formation, the lowest subdivision of the Niagaran limestone of Illinois, should be noted. Bretz (1940) suggested that solution occurred at the bottom of settling shale where it made contact with joints below, filled with water, circulating beneath the shale under artesian pressure. This interpretation of the Joliet depressions may explain other cavities and associated enclosed depressions initiated beneath aquicludes. One such case could be the minor depressions above cavities of the 'potholed surface' beneath lava flows and

Figure 10.6: Small-scale enclosed depressions and associated solution cavities at the 'pot-holed surface' in North Derbyshire, possibly of 'Joliet-type' origin, as outlined by Bretz (1940). 'Toadstone' is a local term for interstratified lavas and sills within the limestone succession. Clayey decomposition products account for their effectiveness as aquicludes. Scales of the solutional features (A) and their broader setting (B) in temporary exposures in the Tunstead quarry face, 4 km east of Buxton, are based on triangulation and levelling on the opposite valleyside.

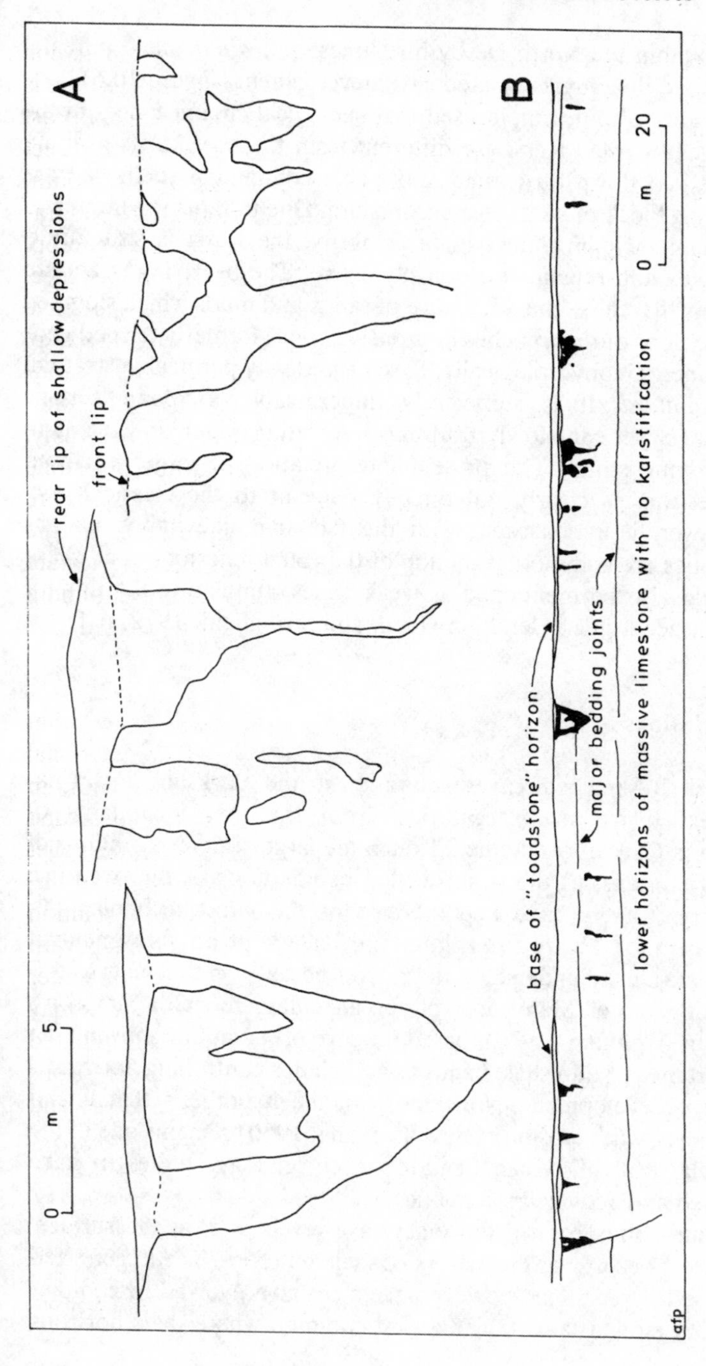
A
0 m 5
rear lip of shallow depressions
front lip
B
base of "toadstone" horizon
major bedding joints
lower horizons of massive limestone with no karstification
0 m 20
afp

sills, within the North Derbyshire limestone, where small phreatic tubes are the only associated cave development (Figure 10.6).

It must also be emphasised that geological circumstances in the classic Yugoslav Karst are different from those of karsts in limestone laid down on a rigid and stable continental shelf, like the Askrigg block of North-west Yorkshire. Due to repeated mountain building, including the Alpine Orogeny, the Karst is structurally complex and remains tectonically active. The overlying strata are usually 'flysch' — the admixture of sands and muds which slumped to the floor of the trenches in orogenic zones formerly termed geosynclines. Geohydrologically, flysch is a slowly permeable stratum, or *aquitard*. It is sufficiently impermeable to pond surface drainage, yet can slowly transmit percolating water downward to underlying strata. The present interpretation of larger enclosed depressions is clearly not directly relevant to the classic Karst. However, it is consistent with the fact that, inevitably, unconformities are common at the top of the Karst limestones and these 'should also be mentioned as weak zones stimulating the flow of subsurface water' (Herak, in Herak and Stringfield, 1972, p. 29).

Conclusions

As briefly sketched, cave exploration in the Yorkshire Dales has proceeded for many centuries, from the first inhabitants of Victoria Cave above Settle, through the large-scale explorations of the Victorians, to the sophisticated explorations of the twentieth century. Cavers have kept abreast of the latest technology to increase their ability to explore, and have kept up with scientific thought so as to understand their caves better — with a view to discovering more! Many now prefer an enlargement of Simpson's view by Waltham (1970), based on cave observations showing the importance of thin shale bands as aquicludes controlling horizontal passage development. None the less, the hypothesis that linked cave levels with erosion cycles (Sweeting, 1950) remains one of the few classically Davisian denudation chronology studies to have reached and activated the public.

Shale bands control the main cave levels and, at the surface, sinkholes occur near the margin of shale outcrops. The larger-sized enclosed depressions, however, seem to have evolved in complementary circumstances, being most common where shale horizons were locally absent. In a broader geomorphological context, therefore, these unusual forms could be grouped with features like 'valley bulges', in which the landform developments within a given stratum are strongly influenced by the character of the overlying stratum. A broader context for the conservation issues outlined can also be envisaged, although these are most apparent in densely populated, long-settled countries like the UK. None the less, a general increase in pressures on the environment is a universally predictable trend. Experience from studying limestone pavements indicates that some unusual landforms and microfeatures can half-vanish before anyone notices, and may continue to disappear or be disfigured even when their uniqueness and irreplaceability are realised.

References

Black, G. (1969) 'Conservation and access', in Cullingford C.H.D. (ed.) *Manual of Caving Techniques* (Routledge & Kegan Paul, London).

Bretz, J.H. (1940), 'Solution cavities in the Joliet limestone of northeastern Illinois', *J. Geol.*, *48*, 337-84.

Clayton, K.M. (1981), 'Explanatory description of the landforms of the Malham area', *Field Stud.*, *5*, 389-423.

Crowther, J. and Pitty, A.F. (1983), 'An index of microrelief roughness, illustrated with examples from tropical karst terrain in West Malaysia', *Revue Geomorph. dyn.*, *32*, 69-74

Duff, P. McL., Hallam, A . and Walton, E.K. (1967), *Cyclic Sedimentation, Dev. Sedim. 10* (Elsevier, Amsterdam).

Dunham, K.C. (1950), 'Lower Carboniferous sedimentation in the Northern Pennines (England)', *Rep. 18th Int. Geol. Cong. Gt Br. 1948*, *4*, 46-63.

Dunham, K.C. and Rose, C.C. (1949), 'Permo-Triassic geology of South Cumberland and Furness', *Proc. Geol. Ass.*, *60*, 11-40.

Ford, D.C. and Drake, J.J. (1982), 'Spatial and temporal variations in karst solution rates: the structure of variability', in Thorn, C.E. (ed.), *Space and time in Geomorphology* (George Allen & Unwin, London and New York), pp. 147-70.

Gams, I. (1974), *Kras. Zgodovinski, naravoslovni in geografski oris* (Slovenska matica, Ljubljana).

Goldie, H.S. (1973), 'The limestone pavements of Craven', *Trans. Cave Res. Gp Gt Br.*, *15*, 175-90.

Goldie, H.S. (1976), 'Limestone pavements, with special reference to North-west England', unpubl. DPhil. thesis, Univ. Oxford.

Green, G.W. and Welch, F.B.A. (1965), 'Geology of the country around Wells and Cheddar', *Mem. Geol. Surv. Gt Br.*

Halliwell, R.A. (1979), 'Gradual changes in the hydrology of the Yorkshire Dales demonstrated by tourist descriptions', *Trans. Br. Cave Res. Ass.*, *6*, 36-40.

Herak, M. and Stringfield, V.T. (eds) (1972), *Karst: Important Karst Regions of the Northern Hemisphere* (Elsevier, Amsterdam).

Hodson, F. and Lewarne, G.C. (1961), 'A mid-Carboniferous (Namurian) basin in parts of the counties of Limerick and Clare, Ireland', *Q. J. Geol. Soc., 117,* 307-32.
Minney, P. (1983), 'Something old, something new', *Homes and Gardens, 64,* 50-55.
Moisley, H.A. (1955), 'Some karstic features in the Malham Tarn district', *Ann. Rep. Coun. Prom. Field Stud. (1953-4),* 33-42.
Moore, D. (1959), 'Role of deltas in the formation of some British Lower Carboniferous cyclothems', *J. Geol., 67,* 522-39.
Phillips, J. (1855, 2nd edn), *The Rivers, Mountains and Sea Coast of Yorkshire,* (John Murray, London).
Pitty, A.F. (1984), *Structure and Relief* (Macmillan Educational, London).
Prestwich, J. (1854), 'On some swallow holes on the chalk hills near Canterbury', *Q. J. Geol. Soc., 10,* 222-4.
Rowell, A.J. and Scanlon, J.E. (1957), 'The relation between the Yoredale series and the Millstone grit on the Askrigg block', *Proc. Yorks. Geol. Soc., 31,* 79-90.
Simpson, E. (1935), 'Notes on the formation of Yorkshire caves and potholes', *Proc. Univ. Bristol Speleol. Soc., 4,* 224-32.
Stanton, W.I. (1982), 'Mendip — pressures on its caves and karst', *Trans. Br. Cave Res. Ass., 9,* 176-83.
Sweeting, M.M. (1950), 'Erosion cycles and limestone caverns in the Ingleborough district', *Geog. J., 115,* 63-78.
Sweeting, M.M. (1955), 'The landforms of north-west County Clare, Ireland', *Trans. Inst. Br. Geog., 21,* 33-49.
Thomas, T.M. (1963), 'Solution subsidence in south-east Carmarthenshire and south-west Breconshire', *Trans. Inst. Br. Geog., 33,* 45-60.
Walsh, P.T. and Brown, E.H. (1971) 'Solution subsidence outliers containing possible Tertiary sediment in north-east Wales', *Geol. J., 7,* 299-320.
Waltham, A.C. (1970), 'Cave development in the limestone of the Ingleborough district', *Geog. J., 136,* 574-85.
Waltham, A.C. (ed.) (1974), *The Limestones and Caves of North-west England* (David & Charles, Newton Abbot).
Ward S.D. and Evans D.F. (1976), 'Conservation assessment of British limestone pavements based on floristic criteria', *Biol. Cons., 9,* 217-33.
Williams, P.W. (1966), 'Limestone pavements with special reference to western Ireland', *Trans. Inst. Br. Geog., 40,* 155-72.

11 VOLCANOES

C.D. Ollier

Introduction

Volcanic activity is the most spectacular of all the processes that shape the earth's surface. It produces some of the most distinctive landforms and provides one of the few instances where large landforms are actually built up rather than carved out by erosion. Erosional landforms characteristically associated with volcanoes are distinctive and differ from landforms produced by the same processes in other settings. Volcanoes are commonly linked with faults and earthquakes, and they may tell us much about the major structural features of the earth.

Unfortunately, because volcanoes are so apparently simple they may be treated too superficially. There are a few stereotypes, their activity and their relationships that are almost caricatures sometimes obscuring more varied and interesting volcanic phenomena. The conical shape, simple radial drainage, the great hazard presented by volcanoes, the fiery girdle of the Pacific, the subduction explanation of volcanic distribution are all examples of elementary and over-simple ideas obscuring a richness of information and interest which is present in the real world.

The physical base

Distribution

A fascinating area of volcanic study and speculation today is the distribution and spatial relations of volcanoes. This is largely because of the revolution in earth science brought about by the concepts of sea-floor spreading and plate tectonics. It is now known beyond reasonable doubt that the sea floors are spreading. Sea floors consist essentially of the volcanic rock basalt, and are quite different from continental masses which consist essentially of granite, with a thin cover of other rocks including sedimentary and metamorphic rocks which actually comprise most of the continental land surface. The basalt of the sea floor can be dated by various methods and consists of stripes of rock of different age. In the North Atlantic, there is a mid-Atlantic ridge with young rocks and some active volcanoes, and the rocks of the sea floor get older towards both North America and Europe. The same situation is found in the South Atlantic, where South America and Africa have moved apart as the South Atlantic appeared between them and then spread by creation of new sea floor in the middle. Continental drift is thus seen as a consequence of break-up of an earlier, larger continent, followed by sea floor spreading.

If the Atlantic is getting wider, one might expect the Pacific should be getting smaller, but the Pacific too is spreading, and new sea floor is being created at an even faster rate than in the Atlantic. Other spreading sites are found between Australia and Antarctica, across the Arctic Sea, and in the Indian Ocean. The creation of new sea floor at spreading sites is often accompanied by volcanic activity, and some volcanoes grow above sea level, such as Iceland, which is virtually a piece of the mid-Atlantic Ridge spreading site that happens to be above sea level.

But if all the oceans are growing,what can happen to keep the earth's volume constant? The answer is *subduction.* This refers to a process whereby old sea floor slides under continental edges, especially around the Pacific. Subduction is generally marked by the presence of deep sea trenches, much earthquake activity, and volcanoes which often erupt violently in contrast to the relatively quiet oceanic volcanoes. They are also different chemically, with more silica than the basalts, possibly resulting from the melting of sediments carried down beneath the crust during subduction. The rocks are typically andesites.

In brief, the plate tectonic hypothesis suggests that the earth is divided into a number of plates, and most of the action, including volcanic activity, takes place at plate boundaries. Boundaries include both spreading sites, such as the mid-ocean ridges, with relatively minor volcanic eruptions, and convergent sites where subduction is thought to operate and where violent andesitic volcanoes are found. There are many minor variations on this theme, and plate tectonics does not account for all the details of volcanic distribution, or all the relationships between volcanic distribution and other features such as plateaus, earthquakes, or non-volcanic arcs, but it tends to be the first assumption in many modern accounts of volcanoes.

The Pacific is ringed by volcanoes as well as having volcanic chains in mid-ocean. Of the approximately 450 volcanoes that have been active in historical times about 350 are in the Pacific hemisphere and, of about 2500 recorded eruptions, over 2000 took place in and around the Pacific. A so-called 'andesite line' can be drawn around the Pacific, demarcating a central area which has only basaltic volcanoes, from the outer area where andesitic volcanoes are dominant, but where basalt may also be erupted. On the western side of the Pacific there is a long line of island arcs where subduction is thought to take place. The arcs have a deep ocean trench on the Pacific side, a line of volcanoes along the arc, and a back-arc basin between the arc and the continent behind. Subduction is thought to take place beneath the arc, but the back-arc basin is itself a spreading site, so some modification of the simple subduction story is required.

Some of the limitations and complications of plate tectonics are described by Ollier (1981) who concentrates on the landscape relating to tectonic setting. For instance, some island arcs such as the Kurile arc between Kamchatka and Japan consist of a single row of active or recently extinct volcanoes behind a trench. Some, like Sumatra, are double arcs with a line of non-volcanic islands between the trench and the volcanic arc. Some arcs are backed by continent (Japan), some by only sea floor (Scotia arc), some are in a back-to-back relationship (Philippines), and some straight line features such as the Tonga-Kermadec rise-and-trench have many features of other arcs so are regarded as 'straight arcs'.

On the eastern side of the Pacific, island arcs are not present, but subduction is thought to take place under the Andes in South America, which has a deep ocean trench offshore, and under the western coast of North America, which lacks a trench. The subduction zone is associated with many earthquakes and andesitic volcanoes. Mt St Helens has produced the most spectacular eruptions on this line recently.

Some other volcanic patterns can be accounted for in different ways. Some volcanoes occur as lines of islands which get younger in one direction. The Hawaiian Islands are a good example, with the present Hawaii as the active volcano. Other islands in the chain, which are extinct volcanoes, become progressively older to the West, showing progressively more erosion, subsidence and coral growth, with drowned valley coasts and fringing coral reefs. Further to the West the volcanic islands are replaced by coral atolls. The Emperor Seamounts, Society Islands, and other island chains show similar trends. In these instances, the sea floor plate may have moved over a 'hot spot', 'fixed' in the mantle below the crust, which caused periodic eruptions and so a chain of islands. The line of islands gives the direction of plate movement, and where the rocks of the volcanoes can be dated the rate of movement of the plate can be calculated. The progressive sinking of volcanoes is due to the heavy load they place on the earth's crust. Like other heavy loads, such as ice-caps, large lakes, or thick piles of sediments, the weight can depress the crust and the volcano sinks under its own weight. Oceanic volcanoes, resting on relatively thin crust, are especially prone to sinking. The situation is complicated by world-wide changes in sea level which is at present generally higher than it has been since the last glaciation. Nevertheless some volcanoes, such as the South Atlantic islands of Gough, Tristan da Cunha and Nightingale have raised beaches indicating relative uplift of the islands.

Not all lines can be explained so easily. For instance the Cameroon line in West Africa consists of Tertiary to recent volcanoes (Mt Cameroon erupted in 1982). These have rift valley 'style', but lack rift faulting. Silicic volcanoes were erupted from 65 million to 10 million years ago, and basaltic rocks have only been erupted in large quantity during the last 10 million years, generally associated with fissures. The line of volcanoes is partly in the ocean (mostly basaltic) and partly on the continent where volcanism was accompanied by uplift of about 1 km. There is no regular trend in the age of the volcanoes along the line. The Andes chain of volcanoes provides a remarkable example of a long chain of central volcanoes, but a double chain appears north of 30° S for no known reason. There is also a gap with no volcanoes between about 8 and 18° S. In North America the lines are less distinct. The Cascade Range of the United States is almost entirely volcanic, consisting of andesites associated with granodiorite intrusions. Many large composite cones are formed including the currently important Mt St Helens. The contemporaneous Columbia River volcanics are basalts, possible related to a spreading site that underlies the western US.

Morphology

Geomorphologists are interested in the shape or morphology of volcanoes, and this aspect is now studied with some mathematical rigour, which fortunately supports and reinforces concepts that were gained by more subjective methods in earlier times. Pike

(1978), for instance, analysed topographic data for 655 volcanoes. Data listed are height, flank width, crater diameter and depth, and circularity. A multivariate statistical analysis suggested that the volcanoes cluster in only seven or eight groups, each of which implies a particular combination of volcano-forming processes. The groups are lava shields, stratovolcanoes (with crater or caldera), cauldron-centred ash-flow plains, small tephra cones, maars, table mountains (tuyas), and silicic lava domes.

Suzuki (1977) analysed the morphology of 699 volcanoes for which sufficient descriptive detail was available, and found they grouped into six 'volcano series'. The six classes and their global population percentages are:

Strato volcano	62
Strato volcano with caldera	10
Shield volcano	11
Shield volcano with caldera	3
Caldera volcano	7
Monogenetic volcano	6

Happily, almost three-quarters of the world's volcanoes conform to the popular image of a volcano — the strato volcano type. Suzuki also divided the world's volcanoes into five regional types: island arc, alpine zone, continental oceanic, and rift system and ridge volcanoes.

By combining information from the morphological classification with that from the regional classification various conclusions can be drawn: lava domes rarely coexist with other kinds of parasitic volcano; about half of the monogenetic volcanoes to be formed in the future can be expected to be lava domes. Some of the results appear to reflect the degree of explosivity, which relates to both volcanic morphology and the global setting. The approach can be carried further, as in Moriya's study of the topography of 142 Japanese volcanoes of a single type, lava domes. He classified them into plug domes, typical domes, dome with flow, flat lava dome, and crypto dome, and further correlated his morphological types with chemical composition and dominant mineralogy. Similarly, Breed (1964) classified scoria cones into five types based on the shape of their crater. Such attention to morphology may seem overzealous, but as Simkin *et al.* (1981) point out different volcanologists have used different terms for the same features and there is still no standardised usage. Geomorphologists could make a contribution here. The catalogue of Simkin *et al.*, uses about 20 morphological volcano types.

Ash, ignimbrite and calderas

Pyroclastic rocks are those built of small fragments of volcanic rock — ash, dust, scoria and bombs. Pyroclastic rocks are especially common amongst the volcanoes of siliceous composition such as andesites and the even more siliceous rhyolites. Individual eruptions can produce widespread, single, often distinctive layers of ash that can be traced over long distances. Such bands are very useful in determining geomorphic history and processes, and in related fields such as archaeology and soil science. Tephrochronology is now an important branch of volcanology. In some places, such as New Zealand, a history can be traced through scores of recognisable tephra layers.

Blong (1982) has provided a fascinating account of how the detailed study of a single tephra layer, the Tibito Tephra of Papua New Guinea, can lead to convincing and important conclusions. Many techniques were used to identify the tephra so it could be mapped, and from the map and its petrology the source could be determined. When the source was known, volumes and energy could be calculated, and theoretical comparison with other volcanoes enabled further conclusions to be drawn. The Tibito Tephra eruption of Long Island took place sometime between 1630 and 1670 and probably lasted two days. The ash volume was $30\,km^3$, the explosive index 99 per cent, and the energy released about 10^{18} joules. It was an eruption comparable with that of Krakatau, and not surprisingly it figures in legends over a wide area of Papua New Guinea.

Another class of volcanic eruption produces ash on a huge scale, with volumes measured in many tens of cubic kilometres, associated with huge eruptions, the formation of vast plains of pumice that blanket pre-existing topography, and with large volcanic depressions called calderas. The ash is produced at great speed and travels as a flowing, glowing cloud or *nuée ardente.* These are ash flow (pyroclastic flow) deposits, in contrast to the simpler ash fall deposits associated with smaller eruptions. A pyroclastic rock composed predominantly of pumice and shards of volcanic glass, showing evidence of features of pyroclastic flow is called an ignimbrite. *Ignimbrites* are often compacted and welded even to the extent that they may look like a solid lava. Ignimbrites may cover thousands of square kilometres.

The eruption of Mt St Helens on 18 May 1980 ejected 0.6 km^3 and left a crater with a diameter of 2 km. Imagine what sort of eruption produced the Yellowstone caldera with a long axis of 70 km, or the Lake Toba caldera with a long dimension of 100 km. The eruption of Taupo, New Zealand, in 186 AD probably provided the most violent volcanic event yet documented. According to Walker (1980) the eruption column rose more than 50 km, carrying 14×10^9 tonnes of pumice into the air, of which less than 20 per cent fell to ground within 200 km of the vent. Erupted ash travelled across the land at 700 km/hr; over 16 000 km^2 of forest were flattened; the ash cloud travelled at least 1350 m above its vent; and the flow was not aided by topography, but passed obstacles up to 1500 m high more than 45 km from the vent.

Hydrology and drainage

Many volcanoes are very porous and permeable, and water soaks through the rock with ease. Not only scoria and volcanic ash, but even many flows are permeable through cracks and joints. Therefore, volcanic areas can be rather arid, at least until they are well weathered, and river courses can be scarce. In the western plains of Victoria, for instance, only one river crosses the volcanic plains to the sea. Groundwater may emerge as springs, which in some areas are very large, and by sapping back into their headwalls they create spectacular landforms such as large alcoves of the Snake River region.

Steep to vertical dykes of intrusive volcanic rocks are often impermeable. If a set of such dykes intrude porous rocks they divide the volcanic mass into a number of separate groundwater compartments, each like a bucket of sand, with the bucket representing the dykes and the sand representing the porous rock in the groundwater compartment. As groundwater accumulates from rainfall the compartment fills to the brim and eventually overflows. At the overflow point a spring will appear, and the stream flowing from it will erode a notch in the dyke. As the notch is eroded down, the watertable in the groundwater compartment will fall. Such a situation exists on the island of St. Helena in the South Atlantic, where a thick dyke behind Jamestown is notched by the overflow channel at the aptly named Heart-Shaped Waterfall. There is abundant groundwater behind the dyke, but when Jamestown last had a drought engineers tried to improve collection of surface water above the waterfall (where the ground is porous) instead of aiming for the deeper groundwater. In more developed countries with larger populations, such as Hawaii, the engineering development of groundwater compartments is well advanced, and tunnels are made through the dykes so that water can be tapped at will with no need for pumping.

In contrast to the Snake River basalts, those of another major lava province, the Deccan Traps of India, are essentially impermeable. There are no scoriaceous surface layers to the flows; the soils are generally rich in impervious clays; of the two main kinds of basalt represented, the amygdaloidal basalts are completely impervious, and the compact basalts, though jointed to some extent, may not contain water because joints are tight at depth, and may be clay-filled or not interconnected. The net result is that percolation is limited and the major portion of precipitation is disposed by run-off. No reliable figures are available, but Gupte estimates that not more than 10 per cent of the precipitation goes into the rocks.

Drainage pattern analysis can be very instructive on volcanic terrain. On volcanic cones the drainage pattern is simple and radial: a beautiful example is shown in Figure 11.1. As an extinct volcano is eroded the radial drainage may be retained and superimposed on the bedrock when this is exposed. Thus even when volcanoes have been largely removed by erosion the drainage pattern may indicate the presence of a former volcano. The volcanic plug can never be wholly removed by erosion, so in investigations where a former cone is postulated the test is to look for the vent, for the radial drainage should radiate not only from the top of the volcano but also from the point of intrusion. In Eastern Australia, where volcanic activity prevailed through the last 90 million years until a few thousand years ago, such analysis is showing that many areas once regarded as simple lava plains actually contain numerous central volcanoes with radial drainage.

Other drainage complications occur when a volcano erupts upon a drainage line, and streams are forced to find a new course around the volcanic pile. Mt Etna is a major example, with the River Alcantara flowing around the north and the River Simento flowing around the south. On Gough Island in the South Atlantic, the existence of several small volcanoes was first worked out from investigation of the drainage pattern. Another odd kind of

Figure 11.1: Vertical air photograph of Mt Egmont, New Zealand (North-West at top of frame). The crater of the strato volcano, partly occupied by a tholoid, is located in the centre of the picture, and from it drainage is dispersed in the classic radial style. The parasitic cone of Fanthem Peak, south-east of the main crater, has its own radial drainage on a smaller scale. To the north-west there is the much dissected remnant of an earlier volcano, Pouakai, but radial drainage is still obvious even though erosion by a stream flowing north-west has cut back and largely destroyed the crater. The slopes of Pouakai intersect the slopes of the younger Mt Egmont, and a typical 'gutter' drainage between two cones runs to the west, and a similar but smaller 'gutter' runs to the north. Some of the south-east drainage of Pouakai has been dammed back by the formation of Mt Egmont, giving rise to a pale-toned patch of alluvium. A parasitic dome that erupted almost on the overflow may have aided the damming process. Similar small parasitic domes can be seen on the southern flank of Mt Egmont. Pouakai, Egmont and Fantham fall on a line, suggesting they are connected to a major fissure at depth. The volcanoes get younger from north-west to south-east.

The almost perfect circle marks the limits of the Egmont National Park, brought out by the contrast between the dark forest and the lighter tones of agricultural land. In a few places the forest extends beyond the National Park, due perhaps to this scene being in New Zealand. In most countries it is more usual to find the agriculture extending into the park.

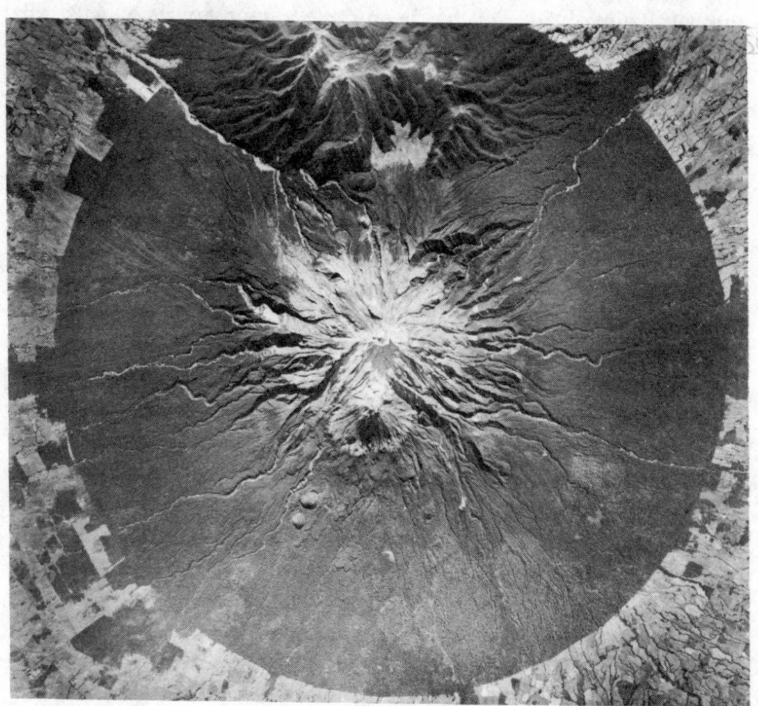

drainage is the 'gutter' created when two volcanic cones intersect (Figure 11.1). In New South Wales, the Upper Hunter appears to be such a valley, formed between ancient Barrington Volcano and Royal Volcano, now both intensely eroded. Many lava flows are tens of kilometres in length, and the Late Quatenary Undara flow in Queensland reaches 160 km. The lava flows fill old valleys, and displace the rivers that once occupied them. In simple situations drainage is carried in lateral streams that run along the edge of the lava flow, but some rivers may be diverted into entirely new courses, or cut across the lava flow. In time, the rivers in their new courses erode new valleys and eventually, there may be complete reversal of relief, with a line of basalt-capped hills marking the line of old valleys. In some regions of old volcanoes the sub-basaltic alluvium contains economic minerals such as gold, tin or sapphires, so the study is also of value in deciphering geomorphic history, for a sequence of the interrelationship of basalt and drainage is of commercial importance. The study of landscape events can be worked out and then often dated by potassium–argon dating of the basalts involved. This has been done in many parts of Australia with surprising results. In many areas the old basalts enable landscape history to be traced back to Eocene times, about 50 million years, with very little apparent change in base level. The basalts also enable quantitative work on such topics as rates of erosion, of weathering, and (more controversially) of tectonic uplift.

The human response

Perception of volcanic hazard

The typical human reaction to volcanoes is one of horror at the great danger and the huge loss of life brought about by volcanic activity. Volcanoes are classed with earthquakes, floods and hurricanes as great natural hazards. How true is this picture? There is no doubt that volcanoes *are* dangerous and do kill people, and such events are newsworthy even if just a few tourists are killed by a very minor eruption as happened on Mt Etna in 1983. But, in general, volcanoes have not proved to be a great natural hazard in human history so far. In the last 500 years volcanoes have killed about 200 000 people, about half of them in just three eruptions — Tamboro, Krakatau and Mont Pelée. This is really a very low toll — very much less than earthquake and flood death tolls, and considerably less than the effects of wars, influenza or road accidents. The one major eruption that had devastating effects on civilisation appears to have been the eruption of Santorini, Crete, c. 1470 BC which destroyed the Minoan civilisation and probably created the legend of Atlantis.

Volcanic hazard is not confined to the direct effect of volcanic action such as being hit, buried or suffocated. Associated tsunamis cause further deaths near the coast; drainage diversions and landslides bring further trouble, and the ruination of crop means that famine often follows a volcanic eruption. The darkness and dreadful din in a prolonged eruption can lead to terror and panic, as in the eruption of Coseguina, Nicaragua, 1835. It was reported that the awful noise and ashfall was taken to indicate the coming of Judgment Day, and the terror of the inhabitants of the town of Aloncho was so great that 300 of those who were living in sin were married at once (Wilcoxson, 1967). In contrast, in St Pierre before the devastating eruption of Mt Pelée which killed 30 000 people in 1902, the city was covered by a fine ash which muted all sounds, and anxiety was aroused by other indicators such as the sulphurous air, dead birds falling from the sky, and the ominous incandescent mass rising in the volcano.

Minor eruptions may actually be beneficial, for plant growth is often encouraged by tephra fall which provides a top dressing fertiliser. Blong (1982) has collated numerous instances in which tephra fall was beneficial. Varenius wrote in 1683 of an eruption of Vesuvius, 'But then the Conflagration ceasing, and the showers watering the Sulphureous *Embers* and *Ashes*, in the Superficies of the Mountain here and there was great fertility of *wine*.' After the 1918 eruption of Katla, Iceland, there was improved grass growth on low-lying lands, and in areas where Paricutin, Mexico, deposited 3 cm of tephra, crops of wheat and barley were excellent. Three years fall was 'the best thing that ever happened to Kodiak'. 'Never was such grass before, so high or as early. No one ever believed that the country could grow so many berries, nor so large, before the ash' (Griggs, 1918).

On Tristan da Cunha, the mid-oceanic volcano that supports 250 people in the remotest permanent settlement on earth, the bulk of the soils are too poor for crops. Agriculture is carried out on a small area of young pyroclastic deposits about 5 km from the settlement, which provides the only fertile land on the island (Figure 11.2). The eruption of 1961 produced cindery lava which

Figure 11.2: The Patches, Tristan da Cunha. Numerous small, parasitic cones are associated with an ash cover that gives the only area of fertile soils used for potato growing. Stone walls are common in volcanic areas, and serve to divide land as stones are cleared from the fields.

is good for making walls but which will remain quite useless for crops for thousands of years.

Blong (1982) has collated many legends associated with an eruption of Long Island, New Guinea sometime between 1630 and 1670, which caused a 'time of darkness' for two or three days. Many people were killed, and not surprisingly the event was commonly considered harmful. However, many benefits were also noted: 'Many pigs and cassowaries were killed at this time, and people ate as if it were a pig feast.' 'Very beneficial — resulted in very fertile land — good crops of everything, but especially sweet potato afterwards.' 'Beneficial — sweet potato came up very strong after darkness. Ash used as "medicine" on sores.' Interestingly the Duna have a legend predicting a future darkness when 'everything will grow gigantically. Men will have big strong children with lots of flesh, nut pandanus and bananas will have big fruit, sweet potato will be big. Despite these perceived benefits Duna don't want it to happen again.'

Although a top-dressing of volcanic ash may be beneficial, deeper burial kills vegetation and leads to famine, and it may take a long period of weathering before the volcanic ash once again provides fertile land. The destructive potential of a volcanic eruption depends mainly on burial. Walker (1980) suggests that burial under 1 m of ash would be destructive, and suggests that the 1 m isopach might be used as a measure for destructiveness. This varies from about 20 to 10 000 km^2 (the Taupo eruption described later is a modest 1000 km^2). This might be a suitable conventional figure for comparing ancient eruptions, but in reality burial by even a few centimetres might be catastrophic to a human population.

Clearly there are harmful and beneficial effects associated with volcanoes. The balance of nature, however, is a concept appreciated mostly by those who don't get eaten. Appreciation of volcanic hazard depends on where you are: an eruption may have some benefits to survivors, but it is nevertheless a personal tragedy for those who get killed. None the less, the potential menace of volcanoes is greater than would be supposed from the disasters that have happened so far. Some ignimbrite deposits are on a tremendous scale, indicating colossal eruptions. Most great eruptions are prehistoric and most of the large historical eruptions, such as the 1912 eruption of Katmai, Alaska, and the 1956 eruption of Bezymianny, Kamchatka, fortunately took place in uninhabited areas. If a comparably huge eruption should occur today in a densely populated area such as Japan, Indonesia, New Zealand or California, there would be a human catastrophe of immense proportions.

Control of volcanoes

Because volcanoes are seen to be hazardous much of the day-to-day work of volcanologists is concerned with monitoring volcanoes in the hope of giving warning of impending eruptions and mitigating their effects. Control of volcanic activity can only be attempted in limited ways, but is an appealing topic for the optimistic.

Attempts to control volcanoes are difficult enough because of the direct physical danger, but are often compounded by social, legal and political problems — it is one thing if God destroys by means of volcanic phenomena; it is quite another if destruction can be attributed to a volcanologist, a geological survey or a politician. An obvious legal problem that arises with flows, for instance, is that the diverted flow will go on to cause damage elsewhere. Who is responsible? Other problems arise from the problematic decision-making process; responsibility for the consequences of decisions such as evacuation; the financing of research, warning, monitoring and control; and insurance. In Hawaii, for example, the owner of a house in the path of an advancing flow will hope that the heat will cause the house to burn down before it is flattened by lava, because he can insure against fire but not against lava. The tragedy of St Pierre, Martinique might have been much reduced if the population had been permitted to evacuate, but the Governor was anxious that they should stay for elections to be held on 10 May. On 6 May troops were stationed to prevent people leaving; the disaster struck on 8 May.

Barriers may be used to divert flows or to dam them. Even if not totally successful they may reduce a potentially long flow into a shorter, thicker one, and so prevent considerable damage. Such barriers were partially effective in Hawaii in 1955 and 1960. During the 1973 eruption of Heimaey, Iceland, barriers were used, as well as high-pressure water hoses to cool the advancing flows, with substantial success. One of the earliest known attempts at control occurred in 1660 on Mt Etna when a lava flow threatened Catania. Fifty men, wearing hides for protection and armed with iron bars, breached a hole in the side of the flow and diverted it. Unfortunately, it then headed for Paterno, and that town sent 500 men to drive away the Catanians. The flow resumed its former course and destroyed a large part of Cantania.

A more complicated scenario involving prolonged conflict between scientists, politicians and a rather irresponsible press followed the 1983 Etna eruption. An attempt to breach a levée on the lava flow by explosives captured most attention, and although it accounted for only 5 per cent of the expenditure on diversion efforts it gained over 95 per cent of media coverage. It was partially successful, but was generally derided by the press as a failure. On their way to the explosive 'experiment' each day the reporters passed within 200 m of a massive construction project aimed at erecting a diversion barrier, which worked well, but had virtually no media coverage. The total cost of all diversional efforts was US $3 million, and prevented losses estimated at about US $25 million (Lockwood, 1983). As Lockwood writes: 'Mitigation of volcanic hazards can now mean more than just evaluation, warning, and evacuation advice; under certain circumstances, volcanologists may also advise on the direct mitigation of hazards through active intervention in ongoing volcanic processes.'

References

Breed, W.J., (1964), 'Morphology and lineation of cinder cones in the San Franciscan volcanic field', *Mus. N. Arizona Bull. 40*, 65-71
Blong, R.J. (1982), *The Time of Darkness* (ANU Press, Canberra).
Griggs, R.F. (1918), 'The recovery of vegetation at Kodiak', *Ohio J. Sci., 19*, 1-57.
Lockwood, J.P. (1983), 'Diversion of lava flows at Mt Etna', *Volcano News, 15*, 4-6.
Moriya, I. (1978) 'Topography of lava domes, Komazawa Chiri' (*Bull. Dep. Geog., Momazawa Univ., Japan*) *14*, 55-69. In Japanese; summary in English, *Volcano News, 1*, 5.
Ollier, C.D. (1981) *Tectonics and Landforms* (Longmans, London).
Pike, R.J. (1978), 'Volcanoes on the planets: some preliminary comparisons of gross topography. Procedures', *Lunar and Planet. Sci. Conf., 9th*, 3239-73.
Suzuki, T. (1977), 'Volcano types and their global population percentages', *Bull. Volcan. Soc. Japan, 22*, 27-40.
Walker, G.P.L. (1980), 'The Taupo Pumice: product of the most powerful known (Ultraplinian) eruption?', *J. Volcan. Geotherm. Res., 8*, 69-94.
Wilcoxson, K. (1967), *Volcanoes* (Cassell, London).

General bibliography

A gazetteer of all the world's volcanoes, and their activity, is:

Simkin, T. *et al.* (1981), *Volcanoes of the World* (Smithsonian Institution, Hutchinson Ross, Stroudsburg).

Standard books on volcanoes include:

Bullard, F.M. (1977), *Volcanoes of the Earth* (Univ. Texas Press, Austin).
Decker, R. and Decker, B. (1981), *Volcanoes* (Freeman, San Francisco).
Francis, P. (1976), *Volcanoes* (Penguin, Harmondsworth).
MacDonald, G.A. (1972), *Volcanoes* (Prentice Hall, Englewood Cliffs).

Books that emphasise the geomorphology of volcanoes are:

Cotton, C.A. (1952), *Volcanoes as Landscape Forms* (Whitcombe & Tombs, Christchurch).
Ollier, C.D. (1969), *Volcanoes* (ANU Press, Canberra).

Leading journals in volcanology are *Bulletin Volcanologique* and *Journal of Volcanology and Geothermal Research*. A lighter but very worthwhile digest of up-to-date information is *Volcano News*, published privately by Chuck Wood, 320 E. Shore Dr., Kemah, TX 77565, USA

12 LANDFORM DEVELOPMENT BY TECTONICS AND DENUDATION

T. Yoshikawa

Views of tectonics and denudation in geomorphology

Landforms are shaped by tectonic movement and sculptured by denudational processes. Davis (1899) deduced landform development by denudational processes, postulating prolonged stillstand of a landmass following rapid uplift. Recognising that the postulate of tectonics simply facilitated the explanatory description of landforms, Davis (1905) mentioned interruptions and episodes of the ideal cycle by tectonic processes, but scarcely considered landform development by concurrent tectonic and denudational processes. On the contrary, Penck (1924, pp. 6-14) believed that landforms were shaped by tectonic and denudational processes proceeding concurrently at different rates, and that the Davisian deduction was only an extreme case of landform development. These two distinctive views of tectonics and denudation in geomorphology have been discussed many times, but their rates have rarely been assessed quantitatively.

Schumm (1963) estimated modern rates of tectonic processes to be about eight times greater than the average maximum rate of denudation, and supported to some extent the Davisian assumption of rapid uplift of a landmass, which allowed little denudational modification of the area during the period of uplift. He suggested the possibility of peneplanation in quiet intervals between orogenic periods. Bloom (1978, pp. 293-4, 303-4) also evaluated rates of denudation to be roughly one-tenth of those of tectonic processes, and inferred the increase in area and altitude of a landmass by uplift likely to be much faster than its lowering by denudation. He considered the Davisian scheme of successive tectonics and denudation to be one of multiple hypotheses in the interpretation of landform development until late Cenozoic tectonic history and climatic change would be better known.

Recent geomorphological study has been mainly directed to denudational processes and their products with little consideration of tectonics. Such a trend in geomorphology may be accepted in tectonically stable areas, where tectonic factors in landform development can be practically ignored. But the circumstances are quite different in tectonically active areas, where recent tectonic processes have so markedly impressed their traces on land surface that landform development cannot be elucidated without regard for a role of tectonics. Tectonic geomorphology has been developed in these areas, clarifying sequences and regional characteristics of tectonic activity. Landform development by concurrent tectonics and denudation, however, has not been investigated intensively even there until recently.

Characteristics of tectonic processes

Contemporary tectonics and its relation to morphogenesis

In the past 100 years Japan has experienced about 100 great earthquakes associated with abrupt crustal deformation, such as uplift and subsidence of extensive areas proportional to their magnitude as well as faulting (Yonekura, 1972). Besides, repeated precise levellings revealed slow and long-lasting vertical displacement of various patterns, the rate of which was mostly 1-2 mm/yr even when no earthquake occurred (Yoshikawa *et al.*, 1981, pp. 16-17).

Recent earthquake faults occurred along former fault lines and their strike-slip displacement was concordant in sense with tectonic movement in the past detected from topographic features of the areas, although their dip-slip displacement was not necessarily so (Kuno, 1936; Matsuda, 1976). Many faults which have been active in the Quaternary have progressively accumulated strike and/or dip-slip displacement (Sugimura and Matsuda, 1965; Okada, 1980) (Figure 12.1). This suggests that earthquake faults have been repeatedly dislocated in nearly similar patterns in recent geological times, and that their activities are one of morphogenetic tectonic processes.

In Japan, vertical crustal displacement was surveyed by repeated levellings at least for three successive periods of 20 to 30 years since the end of the last century. Vertical displacement in wavelength of 20-100 km and longer than 100 km, which corresponded to that of ranges and basins and the width of the main islands, respectively, was progressive or oscillatory in successive periods (Mizoue, 1967). In the areas of progressive displacement, it appears in general that ranges were uplifted and basins subsided

Figure 12.1: Atera fault, Central Japan. The fault runs from the left-hand side of the bridge in the foreground to the upper left. Fluvial terraces in the foreground have been progressively dislocated horizontally and vertically by faulting. The mountains on the right-hand side are the upthrown block tilting rightwards. The upland on the left-hand side is a lava plateau. (Photograph by A. Okada.)

progressively. The areas of oscillatory displacement, however, seem to have influenced more or less by earthquakes in any of three successive periods; abrupt seismic displacement was rather marked in amount and reverse in sense to slow and long-lasting aseismic ones. First-hand data of repeated levellings in short terms, therefore, do not always represent proper trends and rates of morphogenetic tectonic processes. An estimation of morphogenetic tectonics from data of levellings was obtained as follows (Yoshikawa, 1970).

In the eastern part of Shikoku, South-west Japan, the Pacific coastal zone was markedly uplifted, tilting landwards at the time of the great earthquake in 1946 and subsided tilting seaward at the nearly same rate in the pre- and post-seismic periods, whereas the Shikoku Mountains occupying the main part of the island subsided at the time of the earthquake and were upwarped in the pre- and post-seismic periods (Figure 12.2). The Pacific coastal zone has repeatedly experienced major earthquakes at intervals of 100 to

150 years in the past 600 years, associated with crustal displacement almost similar in pattern and amount to that at the time of the 1946 earthquake. Such a characteristic recurrence of major earthquakes suggests the cyclic occurrence of crustal deformation in the similar pattern at intervals of about 120 years on an average. On the basis of this inference, resultant vertical displacement in a seismic cycle of about 120 years can be estimated from the geodetic data in the pre- and co-seismic periods, assuming that the rate of vertical displacement was constant through the interseismic period. The regional distribution of the estimated resultant vertical displacement coincides fairly well in pattern with the tectonic features of landforms in Shikoku (Figure 12.2) and, therefore, is judged to indicate that of morphogenetic tectonic processes.

Figure 12.2: Contemporary vertical displacement (upper) and topographic features (lower) along the levelling route in Shikoku, South-west Japan. K: Kochi, M: Cape Muroto, T: Tadotsu.

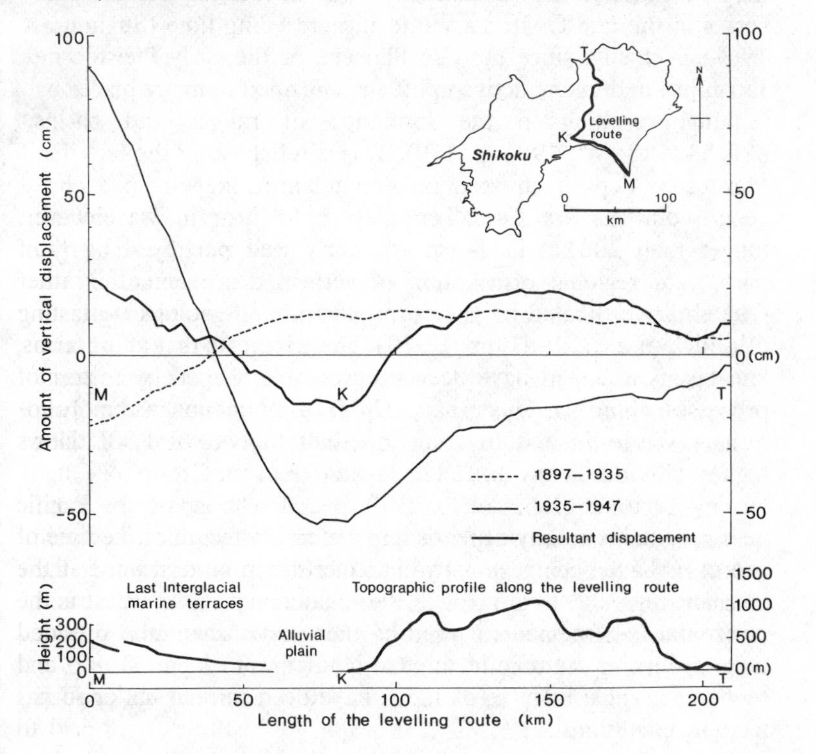

In spite of oscillation in sense and fluctuation in rate, contemporary crustal deformation, when its fluctuating and/or oscillating amounts are integrated for a reasonably long period, is fairly concordant in pattern with tectonic processes as inferred from topographic features. This implies that contemporary tectonic processes, including seismic activities, participate in morphogenesis, and that the landforms of Japan have been developed in recent geological times by tectonic processes similar in pattern to contemporary ones.

Quaternary tectonics as morphogenetic processes

Most of active strike-slip faults have been progressively dislocated in recent geological times. The amount of their displacement, however, increased with time in the Quaternary, but did not in the preceding time (Matsuda, 1976). The growth of Japanese mountains was inferred to have started around the beginning of the Quaternary by extrapolating substantially constant rates of tectonic processes in the late Quaternary into the preceding time (Sugimura, 1967). Actually, since the late Pliocene or the early Pleistocene Japan has undergone active uplift accompanied in many places by faulting, resulting in the formation of ranges and basins (Yoshikawa *et al.*, 1981, p. 119). This is reflected by the fact that Pleistocene deposits are generally abundant in gravel, while Pliocene sediments are fine. Topographic features in wavelength longer than 200 km in Japan are fairly well portrayed by the pattern of regional distribution of vertical displacement in the Quaternary estimated by geomorphological and geological means (Yoshikawa *et al.*, 1981, pp. 17-19). These indicate that mountains and basins in Japan have been progressively shaped by tectonic processes within the Quaternary. Uplift of mountains in the Quaternary was estimated to attain one-half to two-thirds of their highest altitudes in any district of Japan (Research Group for Quaternary Tectonic Map, Tokyo, 1973, p. 33). The growth of Japanese mountains mainly depends on Quaternary tectonics. Tectonic processes have been regionally characteristic in pattern under the constant regional stress through the Quaternary, and vertical and horizontal displacements caused by them have accumulated over time to develop tectonic features of landforms in Japan. The present topographic features of Japan have been formed under such tectonic conditions.

Rates of tectonic processes

Mean uplift rates of the Japanese coast since the last interglacial age were mostly of the order of 10^{-1} mm/yr, but reached higher than 1 mm/yr in tectonically active areas, while those in the Holocene were of the order of 1 mm/yr, attaining 4 mm/yr in maximum (Ota and Yoshikawa, 1978). Mean rates of uplift during the Quaternary were estimated to be 1-2 mm/yr in high mountains of Japan, and mean slip rates of principal active faults in the Quaternary were also of the order of 1 mm/yr (Matsuda, 1976). Mean rates of tectonic processes estimated for various periods often show the tendency that the longer the period the lower the rate. This cannot always be attributed to acceleration of tectonic processes towards recent time, but probably depends on fluctuation in rates of tectonic processes in various periods, as well as inaccuracy in age-determination of old reference features.

Rates of contemporary tectonic processes can be evaluated from data of repeated levellings, by integrating amounts of fluctuating and/or oscillating vertical displacement for a reasonably long period. It is, however, generally hard to estimate for how long these amounts may be integrated. Then, as a result of analysis in the preceding section, the uplift rate of the Shikoku Mountains was calculated to be 1.6 mm/yr in maximum, about a half of the actual rate of uplift in the pre-seismic period (Figure 12.2). The eastern part of the Chugoku Mountains, north of Shikoku, was upwarped during the period preceding the 1946 earthquake, but slightly subsided tilting southwards at the time of the earthquake. The uplift rate of the mountains estimated by the same method as in Shikoku was also approximately a half of the actual one in the pre-seismic period. Vertical displacement in aseismic periods was generally concordant in pattern with topographic features. The uplift rates of Japanese mountains, therefore, were estimated from the actual rates of uplift in aseismic periods, assuming that the ratio of these two rates of 0.5 in the Shikoku and Chugoku Mountains may be applied to the other mountains. The highest rate of contemporary uplift was higher than 2 mm/yr in the Akaishi Range, Central Japan, and 2 mm/yr in the Kyushu Mountains, while it was 0.5-0.8 mm/yr in the Chugoku Mountains.

It is limited to areas along levelling routes where modern uplift rates can be obtained by this method. The estimated rates of contemporary uplift in several mountains traversed by levelling routes, however, are nearly proportional to their maximum amounts of uplift in the Quaternary (Yoshikawa, 1974). The amounts of vertical displacement in the Quaternary, therefore, may be one of the measures of the modern uplift rates over the country. Thus, the modern rate of uplift was evaluated to be 2-3 mm/yr in the high ranges of Central Japan, and around 2 mm/yr in the Outer Zone of South-west Japan, while it was lower than 1 mm/yr in the Inner Zone. It was intermediate in North-east Japan. These rates are of the same order as those estimated for longer periods.

Characteristics of denudational processes

Implications of catastrophic denudational processes in morphogenesis

Precipitation in Japan is characterised by intense rainfall over a short period as well as by high annual amounts. Intense rainfall often causes sudden and large-scale flooding of rivers, especially on the Pacific slope of South-west Japan. Mass-wasting on steep slopes and fluvial action are generally intense, caused by torrential rainfall.

Slope failures, such as debris avalanches and landslips, have played an important role in denudation of Japanese mountains. Among them, debris avalanches most frequently occurred in large numbers on steep slopes of a local extent, triggered by intense rainfall and sometimes by seismic and volcanic activities. The total areas of their scars usually occupied 1-5 per cent of the whole areas, sometimes attaining more than 10 per cent, and the volume of debris slipped down by an accident was mostly estimated to be the order of 10^4 m^3/km^2 (Yoshikawa, 1974). A large number of debris avalanches triggered by seismic shocks occurred densely in 1923 in the Tanzawa Mountains, about 50 km west of Tokyo (Tanaka, 1976). The collapsed area was about 10 per cent of the whole area, and the volume of debris denuded from slopes was evaluated to be the order of 10^4-10^5 m^3/km^2. Since then, their scars have been continuously denuded at an annual rate of 10^3 m^3/km^2. This means that denudation by a single occurrence of densely spread debris avalanches is almost equivalent in volume to subsequent denudation of their scars in about 100 years. It seems that debris avalanches have repeatedly occurred at long intervals in the past at the same places, because minor features of slope sculpture are characterised by their scars formed at different times.

Debris avalanches are one of the most effective denudational processes in Japan.

Mudflows often occur caused by torrential rainfall. Debris transported by mudflows fill valleys and form alluvial cones at mouths of tributaries joining main streams. Mudflows are also a normal denudational process in the upper reaches of Japanese valleys where abundant large boulders are supplied from mountain slopes. The volume of debris transported by mudflows, too, was estimated to be usually the order of 10^4 m^3/km^2 (Yoshikawa, 1974). At present, most Japanese plains are protected by continuous levées from flooding of rivers. Although more or less artificially modified, minor topographic features of Japanese plains are products of rivers freely flooding over them in the past.

Denudational processes in Japan have often caused tragic disasters: people were killed, many houses and public buildings were destroyed, and extensive cultivated lands were buried by coarse sediments. Their occurrence is rare, but occur in all areas. They may be called catastrophic events, but are normal denudational processes in Japan, resulting in noticeable changes of landforms in a short period. This is properly the uniformitarian view of morphogenetic processes, including tectonic ones, such as abrupt crustal deformation associated with earthquakes. To understand the landforms of Japan properly, it is necessary to assess effects of denudational processes on landform development in terms of their magnitude and frequency.

Rates of denudational processes

In Japan, where fragments stripped from steep mountain slopes are generally so coarse as to be transported by rivers chiefly in traction and suspension, sediment delivery rates to reservoirs are one of the measures of denudation rates, because most debris transported by rivers is trapped in reservoirs. It is, however, necessary for this purpose to select reservoirs as follows; (i) those located in the uppermost reaches of rivers with no reservoirs further upstream; (ii) those located in rather narrow valleys, through which debris is apt to be transported continuously; and (iii) those constructed at least ten years — but not long — before, because most Japanese reservoirs are rapidly filled with sediments and then rates of sedimentation decrease with time. The amount of sediment deposited in reservoirs changes considerably year by year due to annual variation of fluvial conditions. It was estimated to be of the order of 10^4 m^3/km^2 in the years when intense rainfall caused extraordinarily high discharge of rivers, while in the years of average fluvial conditions it was of the order of less than 10^3 m^3/km^2 (Yoshikawa, 1974). To estimate mean sediment delivery rates to reservoirs, therefore, it is desirable to take account of the amount of sediments supplied under various fluvial conditions and their return periods.

The mean annual sediment delivery rates to 40 reservoirs in various regions of Japan range from 10 to 10^3 m^3/km^2 in order of magnitude, and are regionally different in a considerable extent (Figure 12.3). The rates in the mountainous areas of Central Japan and on the Pacific slope of South-west Japan are mostly higher than 1000 m^3/km^2, whilst the rates in the Inner Zone of South-west Japan are 20-200 m^3/km^2, and those in North-east Japan are intermediate — 200-1000 m^3/km^2. Assuming that the mean weight of sediments deposited in reservoirs is 1.75 tonne/m^3 and the density of rocks 2.68 g/cm^3, the denudation rate in catchment areas of reservoirs was calculated to be 1-6 mm/yr in Central Japan and on the Pacific slope of South-west Japan, 0.2-0.7 mm/yr in North-east Japan, and 0.02-0.2 mm/yr in the Inner Zone of South-west Japan.

The rate of denudation is generally high in Japan, where catastrophic events play a considerable part in denudation. The estimated rates of denudation in Central Japan and on the Pacific slope of South-west Japan are much higher than those evaluated in the world (Corbel, 1964; Young, 1974); those in the other regions of Japan compare with the latter's. High denudation rates are characteristic in the high mountains of East Asia; for example in the high mountainous area of Taiwan rates were estimated to be 5.5 mm/yr (Li, 1976). This is probably attributed to intense rainfall and its high frequency as well as high altitude and relief.

The actual rates of denudational processes, however, must be higher than the rates estimated from sediment delivery rates to reservoirs, because dissolved materials and a part of fine sediments in suspension are transported further downstream of reservoirs. The annual rate of materials transported in suspension by Japanese rivers was evaluated to be mostly 300-500 tonne/km^2, while the amount of dissolved materials in rivers is probably small due to steepness of landforms in Japan (Yoshikawa, 1974). The actual rates are, therefore, assessed to be at least 0.1-0.2 mm/yr higher than the rates estimated from sediment delivery rates to reservoirs.

Figure 12.3: Mean annual sediment delivery rates to reservoirs in Japan (Yoshikawa, 1974).

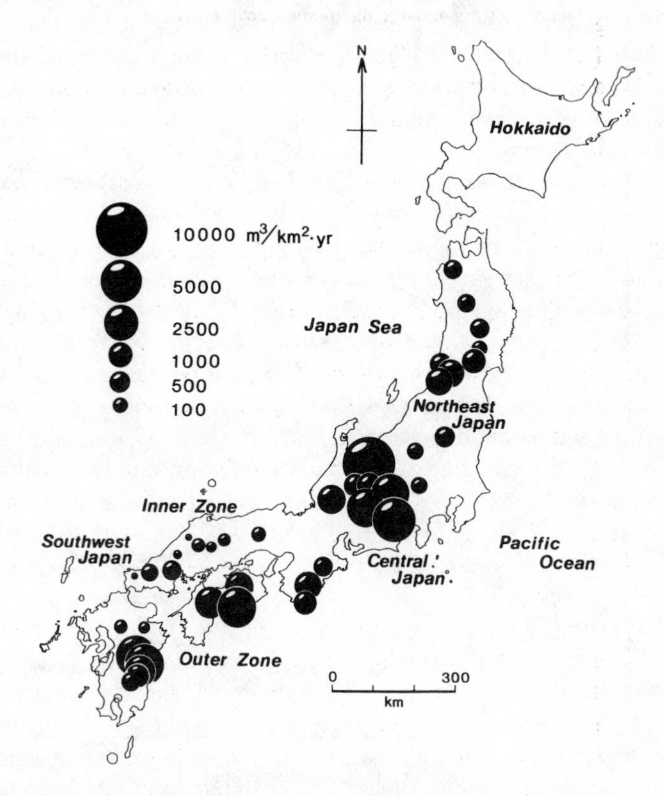

Comparison between rates of tectonic and denudational processes

In general, the present rates of denudation tend to increase with the maximum amounts of Quaternary uplift in the catchment areas of reservoirs (Figure 12.4). This means that the higher the mountains the more intensely they are denuded. The rates of denudation, however, are inferred to depend not only on the height of mountains but also on the other factors, such as relief, lithology, vegetation conditions and the amount and intensity of rainfall, because those to each amount of Quaternary uplift show rather wide range.

As the present uplift rates are nearly proportional to the amounts of Quaternary uplift, the abscissa of Figure 12.4 can be

Figure 12.4: The relation between the amounts of uplift in the Quaternary and the modern rates of denudation in Japan (Yoshikawa, 1974). On the dashed line, the modern denudation rates are equal to the modern uplift rates.

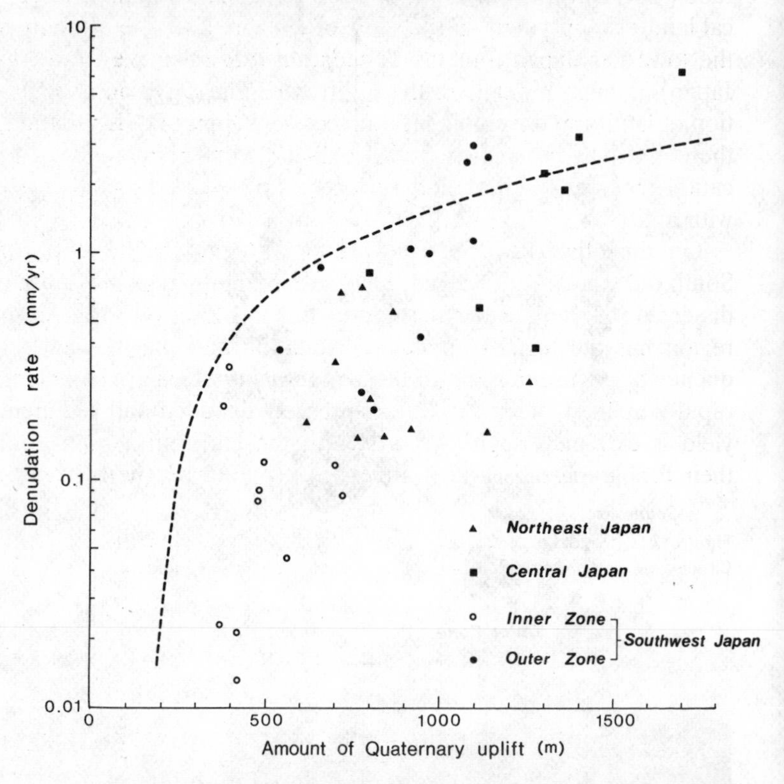

converted into present-day rates. In most catchment areas, the rates are higher than those of denudation. The denudation rates are, however, definitely higher than or nearly equal to the uplift rates in several catchment areas located in Central Japan and on the Pacific slope of South-west Japan, while in the Inner Zone of South-west Japan the denudation rates are much lower than the uplift rates. Such a characteristic regional contrast in the relation between the rates of uplift and denudation is reflected in topographical features of the regions.

The Inner Zone of South-west Japan is a highland region, 400-600 m high, the upper surfaces of which are of low relief and sparsely dissected by narrow gorges. The highland surfaces are erosion surfaces, formed probably in the Pliocene, and cutting the middle Miocene formation as well as the underlying basal rocks. Since then they have been gently upwarped to their present altitude (Yoshikawa *et al.*, 1981, pp. 109-12). The highland is a typical landform of 'youth' in the cycle of erosion. The rate of uplift is the lowest in Japan, but the denudation rate (also the lowest in Japan) is much lower than the uplift rate. The Davisian explanation of landform development by successive uplift and denudation, therefore, may be applied to the highland with the minor modification that weak denudational processes have acted concurrently with uplift.

On the other hand, Central Japan and the Pacific slope of South-west Japan are rugged, high, mountainous regions, densely dissected by deep valleys (Figure 12.5). Moreover, the latter region has the highest annual precipitation and the highest frequency of intense rainfall. In those mountains which are still being rapidly uplifted, steep slopes are intensely denuded and sediment yield is extremely high. According to the Davisian explanation, their topographical features are those of 'maturity'. Their evolu-

Figure 12.5: Hida Range, Central Japan. High peaks are 2800-3000 m above sea level. Cirques were sculptured in the last glacial age. (Photograph by T. Koaze.)

tion, however, cannot be explained by intense denudation following rapid uplift, but by concurrent tectonic and denudation processes, because it seems improbable that the mountains have not been denuded at all, at least during 1 million years, in which time they have been uplifted to their present altitude at the rate estimated from contemporary tectonics. A new framework of landform development by concurrent uplift and denudation, therefore, should be devised to explain their evolution.

Landform development by concurrent tectonics and denudation

A model of landform development in Japan

To elucidate landform development in tectonically active areas, it is essential to investigate changes in the height of mountains by concurrent tectonic and denudational processes. Ohmori (1978) deduced such a change as follows.

Dispersion of altitude. The dispersion of altitude — the standard deviation of the frequency distribution of altitude in a unit area — indicates the roughness of land surface and, therefore, is an index of relief. In Japan, the dispersion of altitude in a certain unit area is regionally different, and in each region it increases with unit area at approximately the same rate over the whole country. This means that the dispersion of altitude in any certain unit area can be used to describe regional characteristics of relief in Japan generally.

The dispersion of altitude increases at a decreasing rate with the mean height of mountains. On the other hand, present-day rates of denudation estimated from sediment yield increase at an increasing rate with dispersion of altitude, which is a measure of potential mobility of materials on land surface. Thus, the relation between present rates of denudation and the mean height of Japanese mountains can be derived as follows:

$$D = 0.3031 \times 10^{-9} \times H^{2.1894}$$

where D is the rate of denudation in m/yr and H the mean height in metres. In other words, the dispersion of altitude becomes increasingly greater by denudation with increase of mean height by uplift, and consequently denudation is intensified.

Critical height in landform development. Change in mean height by concurrent uplift and denudation is expressed as follows:

$$dH/dt = U - D,$$

where t is the time in years and U the rate of uplift in m/yr. Assuming that the rate of uplift is constant and the relation between present rates of denudation and the mean height of mountains is valid over time, the time necessary for mean height to increase from h_1 to h_2 by concurrent uplift and denudation can be calculated by the following formula:

$$t = \int_{h_1}^{h_2} dH/(U - 0.3031 \times 10^{-9} \times H^{2.1894})$$

Figure 12.6 shows the sequential change in mean height of a landmass by concurrent uplift and denudation for each constant rate of uplift. When a landmass is continuously uplifted at a constant rate, its mean height increases with time in the early stage. Change in mean height, however, slows due to intensifying denudation with increase of mean height, and at last the mean height attains a critical altitude that cannot be exceeded by uplift, that is, when the rate of denudation becomes equal to that of uplift. The critical height becomes higher with uplift rate, and the greater the uplift rate the earlier the mountains attain their critical height. Mountains being uplifted at the rate of 3 mm/yr under contemporary denudational conditions reach the critical height of about 1 600 m above sea level in about 2 million years. On the contrary, when uplift ceases, the mean height decreases with time by denudation. High mountains are lowered more rapidly than low ones — it would take about 10 million years for the mean height to decrease to 100 m above sea level, regardless of its initial altitude.

Stage of landform development. Landform development in tectonically active areas can be divided into the following three stages. (i) The developing stage. With increase of mean height by uplift, valleys are deepened by denudation, and the dispersion of altitude becomes greater. The rate of denudation is less than that of uplift, but becomes greater with increase of mean height as well as dispersion of altitude, approaching the rate of uplift. When the rate of uplift decreases or increases in this stage, the mean height increases at a lesser or greater rate than ever, respectively. (ii) The culmina-

Figure 12.6: Sequential changes in mean height of mountains by concurrent uplift and denudation (left) and by denudation (right) (Ohmori, 1978).

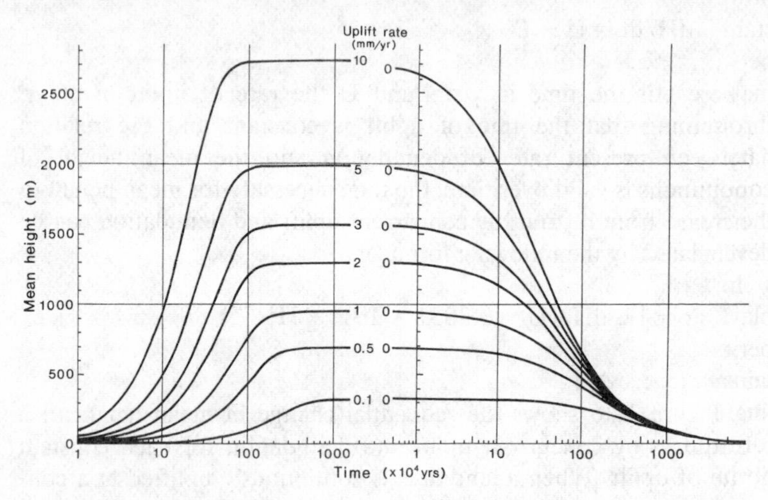

ting stage. When the rates of denudation and uplift become equal, the mean height attains a critical maximum and then remains constant, in spite of continuous uplift. In the higher parts the altitude decreases by more intense denudation than in the lower parts, where the altitude continues to increase because the rate of denudation is less than that of uplift. Consequently, the height of mountains gradually evens to develop the accordance of summit level. The dispersion of altitude becomes nearly uniform over a given area, and the sediment yield remains constant. Thus, landforms of mountains are in steady-state, notwithstanding incessant uplift and denudation. When uplift is intensified or weakened in this stage, the rate of denudation decreases or increases in relation to uplift, and consequently landforms change into a new steady-state at a higher or lower critical height, respectively. While uplift continues, mountains remain in this stage. (iii) The declining stage. When uplift gradually declines or ceases, mean height decreases by denudation, together with the dispersion of altitude and sediment yield. Summit levels are reduced and are more accordant, and landforms indefinitely approach peneplain at extremely slow rates.

Appraisal of the model

The model of landform development in the preceding section postulated the constant rate of uplift and the persistence of the relationship between denudation rates and relief over time. It is, however, still debated in Japan whether the rate of uplift was substantially constant in the Quaternary. Moreover, the relationship between the rate of denudation and relief under contemporary climatic conditions does not seem to have been consistently valid through the Quaternary, because climatic changes in the Quaternary probably caused considerable variation of denudational conditions. Under the present status of our knowledge, therefore, the model may be accepted as a first approximation to landform development in Japan, offering a clue to future study.

In tectonically active and intensely denuded areas, denudation plays a considerable part in landform development during the period of uplift, being intensified with increase of height. Mountains in the developing stage pass through the 'youth' and even the 'maturity' stages in the Davisian sense. Time-independent landforms in steady-state in the culminating stage are in dynamic equilibrium between uplift and denudation (Hack, 1960), while they may be interpreted to be of 'maturity' in the Davisian scheme. Only landforms in the declining stage are within the Davisian realm, but initial topographic features in this stage are not necessarily those assumed by Davis, but those already dissected to a considerable extent. In Central Japan and on the Pacific slope of South-west Japan, where the mountains have been rapidly uplifted in the Quaternary and the rates of denudation compare with or exceed the rates of uplift at present, the rugged high mountains of 'maturity' in the Davisian sense seem to be steady-state landforms at the critical height. The new model is preferable to the Davisian system for the interpretation of landform development in tectonically active areas.

The model is similar to Penck's (1919) analysis of 'die erste Umbildungsreihe' in the development of the accordance of summit level in the Alps. The Penckian view was introduced into Japan, but has not been widely accepted to explain the landforms there, being dominated by the simple and comprehensive Davisian theory. The new model should be further developed and refined to elucidate landform evolution at changing rates and associated development of minor topographic features, clarifying tectonic history and variation of denudational conditions in the late Cenozoic era.

References

Bloom, A.L. (1978), *Geomorphology* (Prentice-Hall, Englewood Cliffs).
Corbel, J. (1964), 'L'érosion terrestre, étude quantitative', *Ann. Géog., 73*, 385-412.
Davis, W.M. (1899), 'The geographical cycle', *Geog. J., 14*, 481-504, in *Geographical Essays* (1909) (Ginn & Co., Boston), pp. 249-78.
Davis, W.M. (1905), 'Complications of the geographical cycle', *Rep. 8th Int. Geog. Cong., Wash., 1904*, 150-163, in *Geographical Essays* (1909) (Ginn & Co., Boston), pp. 279-95.
Hack, J.T. (1960), 'Interpretation of erosional topography in humid temperate regions', *Am. J. Sci., 258–A*, 80-97.
Kuno, H. (1936), 'On the displacement of the Tanna fault since the Pleistocene', *Bull. Earthquake Res. Inst., Tokyo Imp. Univ., 14*, 621-31.
Li, Y.H. (1976), 'Denudation of Taiwan Island since the Pliocene epoch', *Geology, 3*, 105-7, in *Erosion and Sediment Yield* (1982), *Benchmark Papers in Geology, 63* (Hutchinson Ross, Stroudsburg), pp. 335-7.
Matsuda, T. (1976), 'Empirical rules on sense and rates of recent crustal movements', *J. Geod. Soc. Japan, 22*, 252-63.
Mizoue, M. (1967), 'Modes of secular vertical movements of the earth's crust (Part 1)', *Bull. Earthquake Res. Inst. Univ. Tokyo, 45*, 1019-90.
Ohmori, H. (1978), 'Relief structure of the Japanese mountains and their stages in geomorphic development'. *Bull. Dept. Geog., Univ. Tokyo, 10*, 31-85.
Okada, A. (1980), 'Quaternary faulting along the median tectonic line of Southwest Japan', *Median Tectonic Line of Southwest Japan, Mem. Geol. Soc. Japan, 18*, 79-108.
Ota, Y. and Yoshikawa, T. (1978),'Regional characteristics and their geodynamic implications of late Quaternary tectonic movement deduced from deformed former shorelines in Japan'. *J. Phys. Earth, 26 Supp.*, S379-89.
Penck, A. (1919), 'Die Gipfelflur der Alpen', *Sitz. Preuss. Akad. Wiss. Berlin, 17*, 256-68.
Penck, W. (1924), *Die Morphologische Analyse* (J. Engelhorns Nachfolger, Stuttgart).
Research Group for Quaternary Tectonic Map, Tokyo (1973), *Explanatory Text of the Quaternary Tectonic Map of Japan* (Nat. Res. Cen. Disaster Prevention, Tokyo).
Schumm, S.A. (1963), 'The disparity between present rates of denudation and orogeny'. *US Geol. Surv. Pof. Pap., 454-H.*
Sugimura, A. (1967), 'Uniform rates and duration period of Quaternary earth movements in Japan', *J. Geosci., Osaka City Univ., 10*, 25-35.
Sugimura, A. and Matsuda, T. (1965), 'Atera fault and its displacement vector', *Bull. Geol. Soc. Am., 76*, 509-22.
Tanaka, M. (1976), 'Rate of erosion in the Tanzawa Mountains, Central Japan', *Geog. Ann., 58A*, 155-63.
Yonekura, N. (1972), 'A review on seismic crustal deformations in and near Japan', *Bull. Dept. Geog., Univ. Tokyo, 4*, 17-50.
Yoshikawa, T. (1970), 'On the relations between Quaternary tectonic movement and seismic crustal deformation in Japan', *Bull. Dept. Geog., Univ. Tokyo, 2*, 1-24.
Yoshikawa, T. (1974), 'Denudation and tectonic movement in contemporary Japan', *Bull. Dept. Geog., Univ. Tokyo, 6*, 1-14.
Yoshikawa, T., Kaizuka, S. and Ota, Y. (1981), *The Landforms of Japan* (Univ. Tokyo Press, Tokyo).
Young, A. (1974), 'The rate of slope retreat', in *Progress in Geomorphology, Inst. Br. Geog. Spec. Pub.., 7*, 65-78.

13 GEOMORPHIC MEASUREMENTS FROM GROUND-BASED PHOTOGRAPHS

William L. Graf

Introduction

Major objectives of geomorphologic inquiries are the characterisation and explanation of earth surface processes as they vary over time and space. A useful approach to these objectives is the analysis of the changing dimensions of landforms. This chapter outlines methods of measuring heights and horizontal lengths or distances from photographs taken on the ground with hand-held cameras. In remote field locations, accessible only on foot, the photogrammetric approach may represent a cost-effective substitute for more accurate but more expensive techniques. In pilot projects, photogrammetric data may be sufficient to address the issues involved, with more accurate values needed only in more advanced work. However, perhaps the greatest utility of ground-based photographs is their historical dimension. Photographs of a particular geomorphic setting taken several decades ago show scenes and yield dimensions of landforms which differ from those of the present day (Malde, 1973). The former dimensions can be determined photogrammetrically from the historical view, the scene relocated in the field, and present-day dimensions of the features resurveyed in the field. Exact relocation of the original camera position and duplication of the historical photograph with a modern image (Figure 13.1) provides striking evidence of the progress of geomorphological processes (Graf, 1978, 1979).

Sources of historical photography

The variety and obscurity of sources for historical geomorphology photography brings about a convergence of the interests of the geomorphologist, the historian and the detective. Obvious sources include the photographs taken by geographical and geological surveyors who were sometimes more interested in the rock strata than in the landforms shown. Government agencies responsible for the survey reports frequently retain voluminous files of unpublished photographs that are available to those willing to examine the collections, one photograph at a time, in a search for useful material. Local historical societies frequently maintain collections of photos that are geographically indexed, if not also by subject. Subject indices, however, may be unhelpful because only one feature of the photo may be listed even though the image shows other unlisted features of interest to the geomorphologist. Private collections are also valuable and frequently repay the extensive enquiries required to locate them.

Figure 13.1 shows a typical example of a historical photograph and its modern relocated counterpart. The original photograph was taken in 1941 as part of an engineer's flood survey effort and was stored unmarked and unindexed in a government file. The modern photograph was taken in 1982. Figure 13.1 is a representative case for the following discussions of photogrammetric techniques, but the techniques are equally applicable to similar photographs regardless of their setting.

Components of the ground photograph

As with most specialties, *photogrammetry* (the process of making quantitative measurements from photograph-like images) has a set of specialised terms. Concepts employed in this chapter include: focal length, depression angle, principal point, x-axis, y-axis, true horizon, and vanishing-point. Focal length refers to the distance from the centre of the camera lens to the centre of the film plane in the camera. In modern, hand-held cameras this distance is usually in the range of 28-55 mm, but in older cameras it is highly variable. In some nineteenth-century cameras, focal lengths of 50 cm or more were common. Usually focal length (f) must be determined from a known dimension (D) and the image of the same dimension produced by the camera (d). The known dimension must be orthogonal to the line of sight (i.e. directly across the view of the photo-

Figure 13.1: An example of a historical and a modern relocated photograph of the same view: the central highway bridge over the San Francisco River at Clifton, Arizona. Upper view, 31 March 1941, United States Department of the Army, Corps of Engineers, Phoenix Office, Gila River Flood Survey File Photo no. 59A. Lower view, 14 September 1982, by the author.

graph). If the distance (F) from the camera to the object being measured is known (either measured from a large-scale map or preferably measured in the field by relocating the original camera station), the focal length of the original camera can be calculated as:

$$f = (F \times d)/D \tag{1}$$

By way of example, if the camera-to-object distance is 20 m, the dimension of the object on the photograph is 10 mm, and the actual dimension of the object is measured in the field as 4 m, equation (1) yields a focal length of 50 mm, the standard 50 mm length found in many modern cameras.

If measurements are made of dimensions from prints instead of original negative images, an apparent focal length must be determined because the print is frequently not the same size as the original negative which governs all the geometric assumptions for photogrammetric analysis. This apparent focal length can be determined using equation (1) and a known dimension shown on the image, and can be used in all subsequent calculations pertaining to the print.

Figure 13.2 shows the basic geometry of a typical photograph taken with a hand-held, ground-based camera. The geometry is comparable to the oblique aerial photograph (American Society of Photogrammetry, 1960, pp. 25-48, 875-918). The depression angle (θ) is formed by the angle between a horizontal plane and the line drawn from the centre of the film through the centre of the camera lens: it corresponds to the dip-angle familiar to mappers of geologic strata. In many cases the depression angle can be measured with an inclinometer or Brunton compass. The depression angle can also be determined from measurements made on the photographic images using the principal point, y-axis, and the true horizon.

The exact centre of the frame on the film or its printed image is the principal point of the photograph through which pass a vertical line (the y-axis) and a horizontal line (the x-axis). Real parallel lines directly away from the camera position will eventually converge when they appear to meet the y-axis at the vanishing-point. The true horizon is a horizontal line on the photographic image that crosses the y-axis at the vanishing-point. It is frequently con-

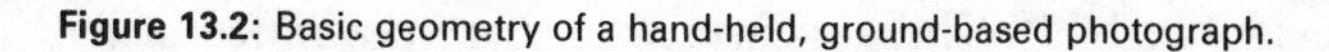
Figure 13.2: Basic geometry of a hand-held, ground-based photograph.

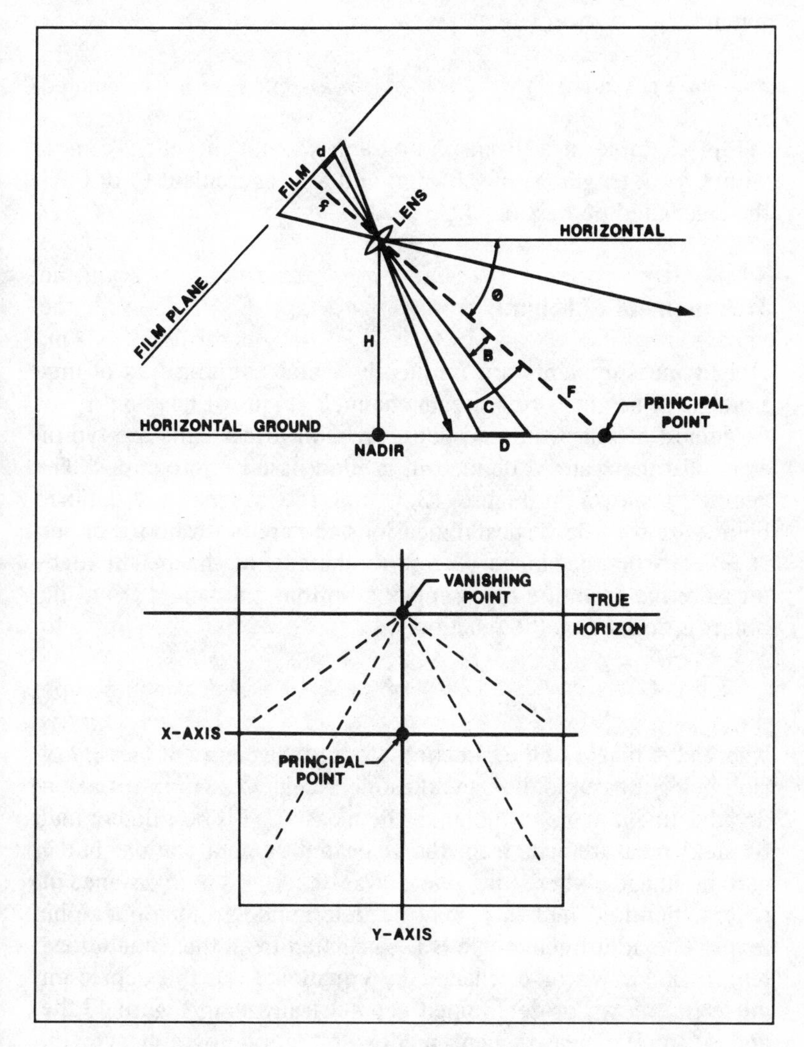

venient to inscribe these features on a copy of the image to be analysed before proceeding with measurements.

The depression angle (θ) can be calculated directly from a photograph that has a defined true horizon, a measured dimension tp (which is defined as the distance along the y-axis from the vanishing-point on the true horizon to the principal point), and a known focal length (f). These measurements are combined in the function

$$\theta = \text{cotan}\,(tp/f) \tag{2}$$

As an example, if a given photograph was taken with a camera with a focal length (f) of 0.05 mm and a measurement tp of 0.03, the calculated depression angle is 31°.

Measurement of heights

Height measurements are frequently useful as indicators of erosion. Bank heights along stream channels (Figure 13.1), in arroyos or gullies, at scarps in landslides, or in wave-cut cliffs are typical examples that can be defined in ground-based photographs. The geometry shown in Figure 13.3, and the calculations outlined below, are for true vertical dimensions and are not valid for slopes. If point b on the image represents the top of the height to be measured and point c represents the bottom, the height (h) of the object is defined by the function:

$$h = H\left(1 - \frac{\tan b}{\tan c}\right) \tag{3}$$

where H = height of the camera bottom to the centre of the dimension being measured, b = an adjustment angle for point b, and c = an adjustment angle for point c. The measure H is best determined by field measurement from the relocated camera station, but in certain images where the photo was taken from a distance of several hundred metres it may be determined from topographic maps. The adjustment angle B is calculated from the camera focal length (f, known or calculated by equation (1)), the depression angle (θ, known or determined geometrically as in Figure 13.2), and a simple measurement made on the photograph (yb, the distance on the image from the x-axis to point b measured parallel to the y-axis). The functions for adjustment angle B are:

$$B = \theta \pm b \tag{4}$$

where the appropriate sign is + if the point b falls below the x-axis and − if b falls above the x-axis. The function for b is

$$b = \text{cotan}\,(yb/f) \tag{5}$$

Similarly, the adjustment angle C is calculated from the camera focal length (f), the depression angle (θ), and a simple measurement made on the photograph (yc, the distance on the image from the x-axis to point c measured parallel to the y-axis; see Figure 13.3). The function for adjustment angle C is:

$$C = \theta \pm c \tag{6}$$

where the appropriate sign is + if the point c falls below the x-axis and − if b falls above the x-axis. The function for c is

$$c = \text{cotan}\,(yc/f) \tag{7}$$

As an example, assume that Figure 13.3 shows line bc which is the height of a bank in a dry stream. The focal length (f) of the camera is 0.05 m (so expressed instead of in mm in order to maintain consistent units throughout the calculations). The height (H) of the camera above point c is 5 m. The depression angle (θ) is 30° (from field measures), yb is +0.01 m (measured from the photograph in consistent units), and yc is +0.02 m (again measured from the photograph). These inputs reveal an adjustment angle B of 41.3°, adjustment angle C of 51.8° and a calculated height of 1.5 m.

Measurement of ground lengths

Measurement of ground lengths entails the analysis of dimensions that are horizontal and are of two types, each with a separate set of procedures. Parallel ground lengths are those that are parallel to the x-axis on the image, that is, across the line of sight. Diagonal ground lengths are not parallel to the x-axis and so have a variety of orientations. Ground lengths are most accurately determined if they occur near the y-axis of the image, close to the centre of the frame. Distortion increases away from the y-axis toward maximum amounts at the sides of the frame, but such distortion is not geomorphologically significant unless wide-angle lenses of focal lengths shorter than 28 mm are used.

Figure 13.3: Geometry for measurement of heights.

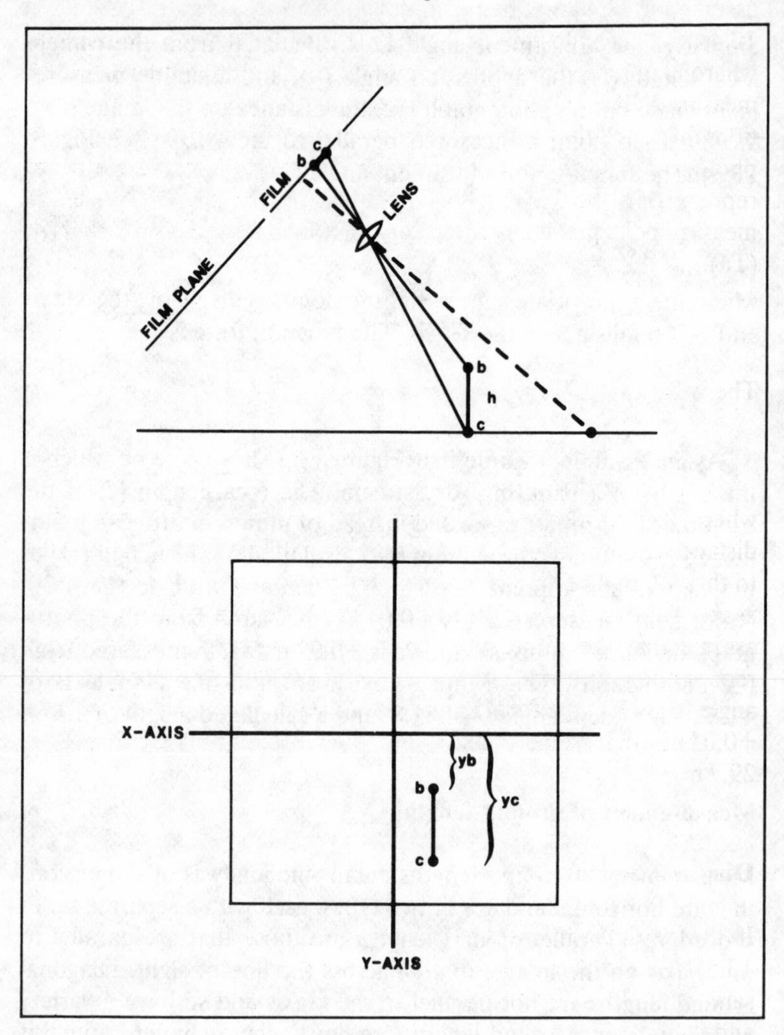

Parallel lengths

Parallel ground lengths are of interest to the geomorphologist in measuring channel widths (Figure 13.1), dimensions of joint spacing in rock faces, sizes of channel bars, and similar dimensions. Parallel lengths measured by the technique outlined below are those horizontal to the imaginary surface of the earth, which is assumed to be flat in the restricted area shown in the image, Figure 13.4, and the calculations outlined below depend on a geometry whereby point a represents the ending point of the dimension to be measured. The length of the dimension (D) is defined by a function that includes a measurement of the dimension (d) on the photograph, height (H) of the camera above the horizontal surface representing the ground level, the depression angle (θ), and a measurement made on the photo pertaining to the true horizon (Ta):

$$D = \frac{d\,H}{Ta\ Cos} \qquad (8)$$

The value of Ta is given by the function

$$Ta = f\tan\theta \pm ya \qquad (9)$$

where f = focal distance, θ = depression angle, and ya is the distance on the image from the x-axis to point a measured parallel to the y-axis. The sign in equation (8) is + if point a falls below the x-axis is the image, − if point a falls above the x-axis. For example, assume that the dimension (d) in a photograph is 0.02 m long, representing the width of a channel at point a. If the depression angle (θ) is 30°, the focal length (f) is 0.05 m, and the measure ya is +0.01 m (below the x-axis), the dimension in question (D) is 29.7 m.

Diagonal lengths

If the desired dimension is oriented other than parallel (Figure 13.5), the geometric constructions shown in Figure 13.6 are necessary, where the dimension in question is the line between points a and b. Two construction lines are required: the first a line from the vanishing-point through point b and extended beyond; the second a line parallel to the x-axis through point a and extended to an intersection with the first constructed line at newly defined point c (Figure 13.6).

Calculation of the length of line ab depends on three steps: (i) calculation of the length of line ac; (ii) calculation of the length of line bc; and (iii) use of Pythagorus' theorem and the lengths of ac

Figure 13.4: Geometry for measurement of parallel ground lengths.

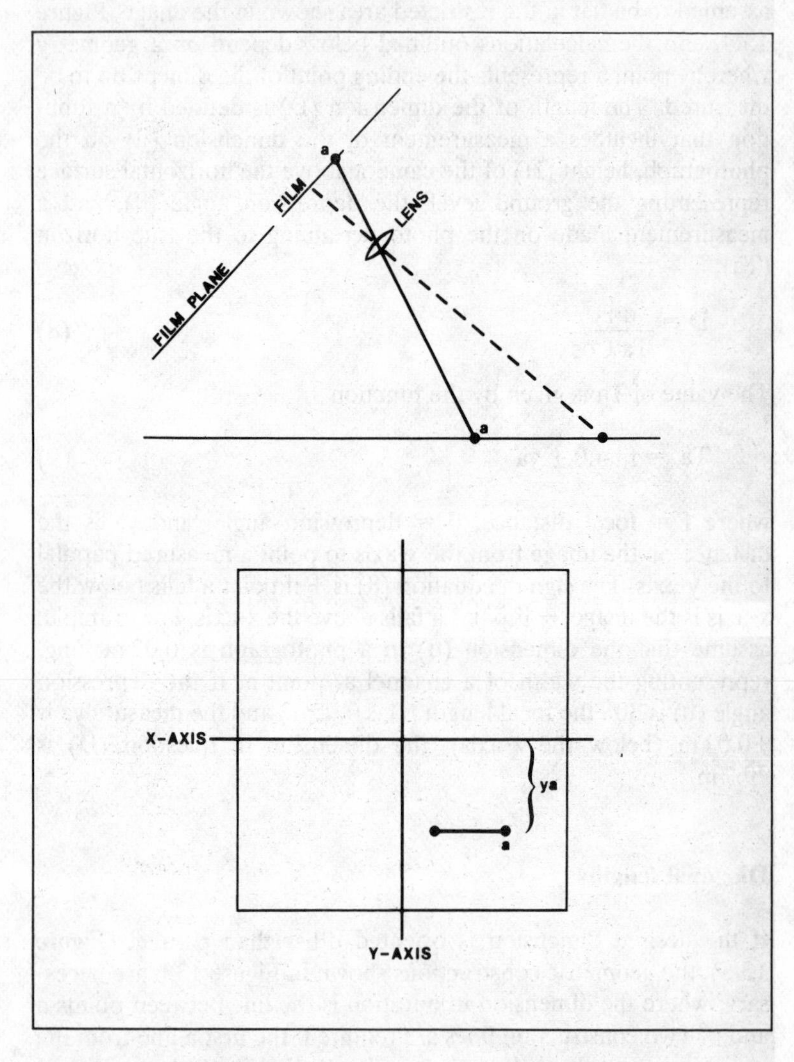

Figure 13.5: An example of a pair of images on which a diagonal ground measure would be of interest: a line defining channel width orthogonal to the direction of flow and located to the right of the bridges. The views show the San Francisco River at the southern highway and rail crossings in Clifton, Arizona. Upper view, *c.* 1937, United States Department of the Army, Corps of Engineers, Gila Flood Survey File Photo no. 64. Lower view, 15 September 1982, by the author.

Birds' Eye View
South Clifton
Clifton Arizona

Figure 13.6: Geometry for measurement of diagonal ground lengths.

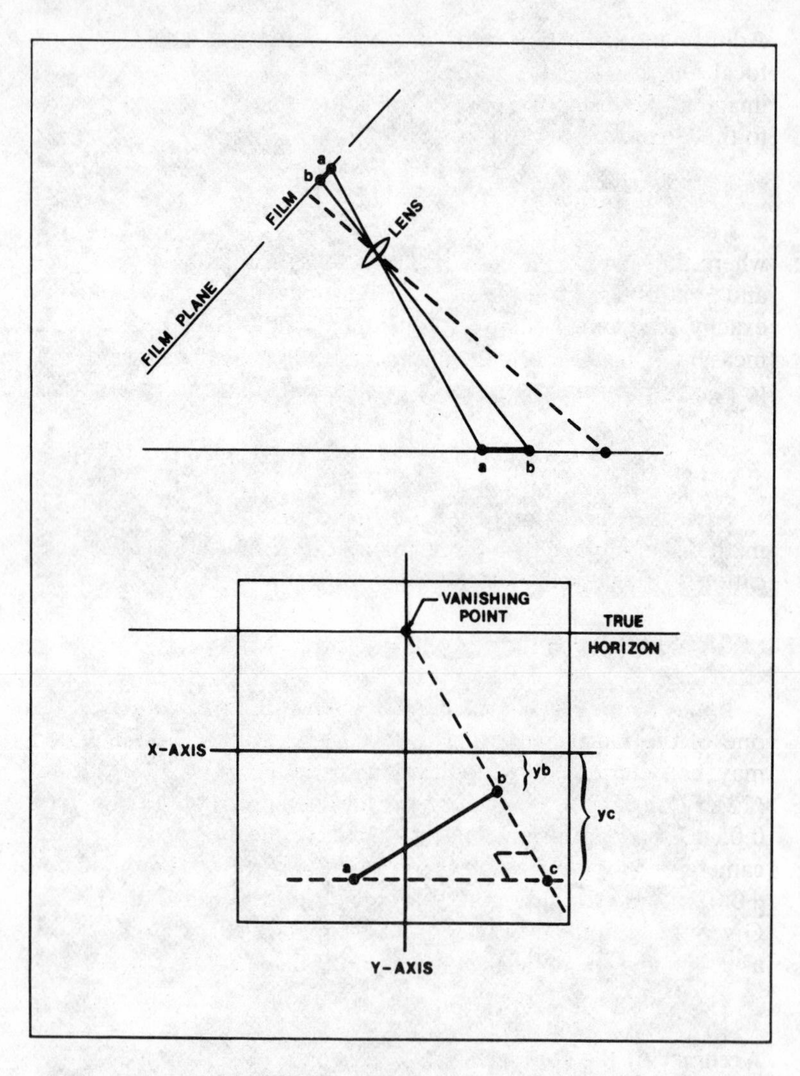

and bc to calculate the length of bc. First, the length of line ac is a simple matter of the solution of the parallel length problem as outlined in the previous section. Second, determination of the length of line bc relies on a function that includes the height (H) of the camera above the horizontal plane representing the ground surface and two adjustment angles, B and C:

$$bc = H\,(\text{cotan}\,B - \text{cotan}\,C) \tag{10}$$

Adjustment angle B is defined by the depression angle (θ) the focal length (f) of the camera, and the measurement yb on the image as the distance from the x-axis to point b measured parallel to the y-axis:

$$B = \theta \pm \text{cotan}\,(yb/f) \tag{11}$$

where the sign is + if point b falls below the x-axis on the image and − if it falls above the x-axis. Adjustment angle C is defined in exactly the same fashion as adjustment angle B except that the measure yc is used. It is the distance on the image from the x-axis to point c measured along the y-axis and included in the function:

$$C = \theta \pm \text{cotan}\,(yc/f) \tag{12}$$

After the calculation of the lengths for ac and bc, the third step in the determination of the diagonal dimension ab is the application of Pythagoras' theorem:

$$ab = \sqrt{(ac)^2 + (bc)^2} \tag{13}$$

It may happen by chance that the unknown dimension lies along one of the radial lines from the vanishing-point. The dimension may be determined by application of equations (10), (11) and (12). For example, assume that the focal length (f) of the camera = 0.05 m, the depression angle (θ) = 30°, the height (H) of the camera = 5 m, the measure ya = +0.02, yb = +0.01 m and yc = +0.02 m (ya, yb and yc all fall below the x-axis and so are +). Given these inputs the dimension bc can be calculated as 1.4 m, ac as 2.4 m, and the dimension in question, ab, as 2.8 m.

Accuracy of the approach

The results of photogrammetric techniques on images from hand-held, ground-based cameras are accurate to within 5 per cent of dimensions measured by tape. The level of accuracy equals that of reconnaissance surveys with unsophisticated instruments but does not, of course, match the precision of on-ground surveys with theodolite methods. Inaccuracies are mainly due to the distortion of the image and to the unstable physical base of the image. All photographic lenses have some distortion due to imperfections in the glass, and hand-held cameras may not require the quality of lens material and construction of aerial cameras. The photographic image used by the analyst is usually a print on paper which expands and contracts with changing temperature and humidity. Historical photographs on paper are especially susceptible to deterioration. However, photographs printed and stored on metal or glass plates, a technique common in the late nineteenth century, produce better, more stable images for analysis.

Detailed measurements are difficult to make from paper or negative images, even with an engineer's scale. Such measurements, and reduction of their inaccuracies, are easier with a photo interpreter's reticule, that is, a tube with a magnifying glass at one end and a transparent target at the other. When the observer's eye is placed over the magnifying glass and the target is placed over the portion of the image to be measured, a scale etched into the target appears superimposed over the photograph. If available, a digitiser is equally accurate.

Particular difficulties in the measurement process include the determination of the measure H (the height of the lens above the horizontal surface representing the ground). If the dimension being measured is several hundred metres distant and the view is taken from the edge of a cliff, for example, detailed topographic maps provide a useful value. For those images taken by a photographer standing on a surface that also appears in the foreground of the photograph, and with the dimension in question a few tens of metres distant, the estimation of H is more difficult. Other complications include the determination of the depression angle and geometric constructions on the photographic image needed to make certain measurements, such as diagonal ground lengths.

Conclusions

Where available, ground-based images made by hand-held cameras offer a viable and economical source of quantitative data for landform dimensions. In some situations they provide the only source of information. Such photographs became available from the 1850s onwards. The approach therefore extends the time-span beyond that covered by aerial photography, which generally did not begin prior to 1925. The scale of the approach also complements the general overviews of aerial photography, for which scales in the 1:20 000 to 1:60 000 range are common, by providing reliable measurements in the 0.1 to 100 m range. This greater detail focuses on scales of many features, such as stream bank heights, small channel widths, and landslip profiles, which are of particular geomorphological interest.

Some inevitable approximations and difficulties in the analysis have been identified. However, only simple measurement tools and a hand calculator are needed to solve relatively simple trigonometric functions which yield measurements sufficiently accurate for most geomorphic purposes. The opportunities of photogrammetric analysis of historical images, followed by relocation and resurvey, permit long-term, process-oriented studies that would otherwise not be possible. Thus, ground-based photogrammetric analysis can extend geomorphological studies in time as well as in space.

References

American Society of Photogrammetry (1960), *Manual of Photogrammetry* (Am. Soc. Photogr., Falls Church, Virginia).

Faig, I.W. (1976), 'Photogrammetric potentials of non-metric cameras', *Photogr. Engng. Rem. Sens., 42*, 47-69.

Graf, W.L. (1978), 'Fluvial adjustments to the spread of tamarisk in the Colorado Plateau Region', *Bull. Geol. Soc. Am., 89*, 1491-501.

Graf, W.L. (1979), 'Mining and channel response', *Ann. Ass. Am. Geog., 69*, 262-75.

Malde, H.E. (1973) 'Geologic bench marks by terrestrial photography', *US Geol. Surv. J. Res., 1*, 193-206.

14 CITIES AND GEOMORPHOLOGY

Ian Douglas

Land surfaces changed by urban development, the paving of the surface, artificial drains and altered albedo have a modified hydrological cycle. Soil erosion may be particularly high during construction and streams may adjust to altered water and sediment flows. New landforms are created to improve the appearance of the city, and by extraction of building materials, earthworks for traffic routes and the dumping of waste. Careful evaluation of the geomorphology of urban areas and assessment of the suitability of the land for different types of building, density of housing and open-space uses can reduce construction costs and avoids on-site and off-site damage. Therefore, the application of an understanding of processes and the evolution of landforms to urban areas has been a recent and rapidly expanding field of practical geomorphology.

In this chapter, the impact of urbanisation on geomorphic processes is examined in relation to rivers, glacial deposits, subsidence, karst, beaches and expansive soils. In particular, the problems arising from the intense rainfalls and deep weathering profiles of most low-latitude cities are emphasised.

Cities and rivers

Control of river flooding by building levée banks, a process often termed channelisation, is widely used. However, regularising the channel, although accelerating river flow, reduces entrainment of sediment. Such channel stability is an effective protection, but may increase erosion downstream. For example, the floodplain of the River Mersey has been changed by the construction of the M63 motorway which crosses the river four times between Stockport and Urmston, near Manchester. Here, the river channel has been embanked throughout, and is regulated by flood basins at Didsbury, Chorlton and Sale. North of Gatley, part of a meander was cut off and the channel shortened. At Northenden, culverts trans-

Table 14.1: Change in sinuosity* of River Mersey from Crossford Bridge, Sale, to Wheat Hey, Ashton-upon-Mersey, 1945-71

Date	Sinuosity*
1845	1.77
1904	1.68
1927	1.67
1954	1.50
1971	1.28

*Sinuosity is channel length divided by reach length.

fer water across the flood plain from the upstream to the downstream side of the motorway, where its return to the river channel is regulated by flood-gates.

At Chorlton and Sale, the gravel pits, used to supply aggregates for the motorway works, have been converted into recreational lakes or water parks. The lakes also store floodwaters diverted from the river by lifting offtake weir-gates upstream of the water parks. The long history of modifications, culminating in the M63 constructions, has reduced sinuosity (Table 14.1). Several meanders have been cut off and reaches straightened (Figure 14.1). Such changes increase the hydraulic efficiency of the channel, maintaining adequate depth and velocity to evacuate all but the severest flood flows without spillage.

However, such river training works can lead to dramatic channel adjustments downstream. The newest training works (Figure 14.1) end 0.5 km upstream of Old Eea Brook at Urmston. Immediately downstream, the river bends with a radius of approximately 50 m, about twice the channel width. Severe bank erosion and channel migration occurred in the period 1979-84, with the stream reducing its channel length and tending to straighten its course. Such bank erosion, with rotational slumps in the stratified flood plain sediments, is typical of many regulated rivers. Stable channels in alluvium require that slope is not steepened by shortening the river course drastically and that channel bends are maintained (Richards, 1982). On the Mississippi severe bank attack occurs if the bend radius is less than 2.5 times the channel width.

Such channel changes are widespread on the Pennine streams which debouch on to the Cheshire–Lancashire plain. Sometimes a change in shape may depend on major floods which shift channels.

Figure 14.1: The changing course of the River Mersey south-west of Manchester between Urmston and Ashton-upon-Mersey, showing the limits of the most recent channelisation works.

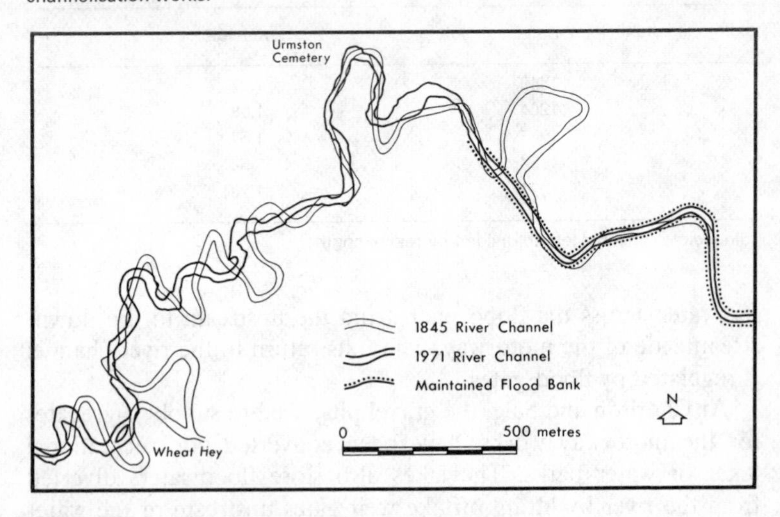

The River Bollin channel was stable between 1872 and 1935. Later, channel instability and meander cut-offs decreased sinuosity from 2.34 to 1.37 (Mosley, 1975). This change in shape was triggered by large floods in the 1930s, possibly related to higher peak discharges due to agricultural land drainage and urban development.

Urban channel adjustment in the humid tropics

In the humid tropics, the peak run-off per unit area in the floods is much greater. On the Sungai Gombak in Kuala Lumpur, Malaysia, peak discharges from the 140 km^2 catchment above Jalan Tun Razak reach 3.1 $m^3 \ km^{-2} \ s^{-1}$. Rapid land clearance for housing developments exposes much deeply weathered ground surface material to the erosive impact of falling raindrops. Rills and gullies develop rapidly on the bare ground, carrying large volumes of sediment. Eventually, the main streams change their shape to cope with these increased sediment and water imputs.

Such a metamorphosis, from a meandering to a braided state, has occurred on the Sungai Anak Ayer Batu (Figure 14.2; Table 14.2) (Douglas, 1975, 1978). Under forest conditions, the Anak Ayer Batu would have been a narrow, meandering stream with a relatively low sediment load. It would have had few tributaries, but

Figure 14.2: Channel metamorphosis during urban development on the Sungai Anak Ayer Batu, Kuala Lumpur.

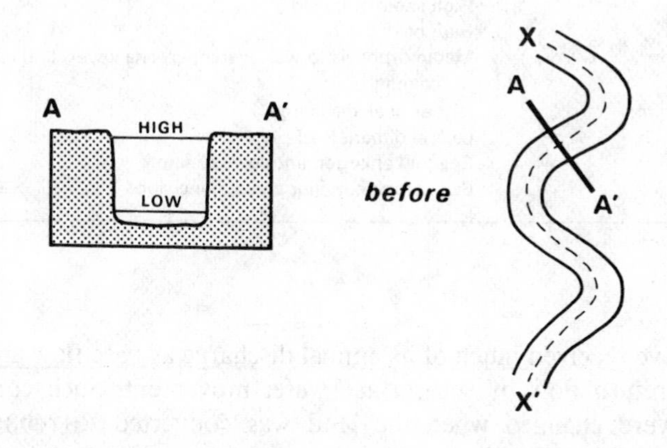

cross section

plan

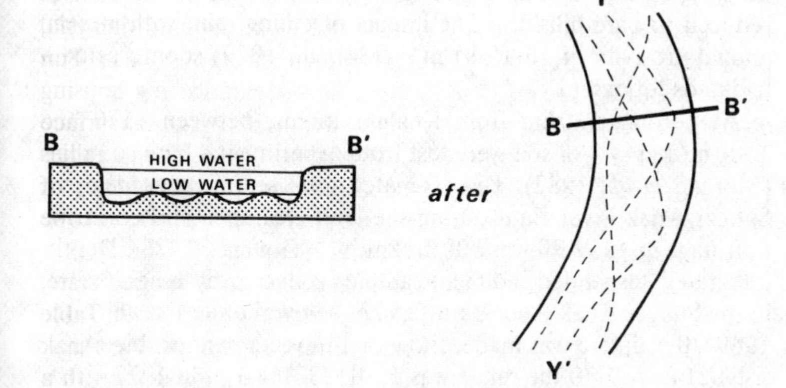

Table 14.2: Sequence of events Sungai Anak Ayer Batu

Forest:	Narrow, meandering Low sediment load
Rubber plantation:	Gullying during clear weeding Peak discharge increased Channel slightly widened Later stabilised, few cut-offs
Construction:	High sediment yield High peak discharge Metamorphosis to wider, steeper, shallower, braided channel
Channelisation:	Higher peak discharge Less sediment load Channel enlargement downstream Bank erosion, minor channel incision

would have received much of its annual discharge as base flow and delayed return flow by sub-surface water movement. Such conditions were changed when the land was converted to rubber plantations. Clear weeding at the turn of the century left the soil poorly protected and gullies developed. Run-off during storms became more rapid and flood peaks higher. Even after the introduction of ground-cover crops, the extended drainage network continued to function but without the severe soil loss associated with the initial gullying. Decades later, the rubber plantations were cleared for housing development. Extensive areas, such as the Bangsar Heights and Damansara Heights private housing estates, were reduced to bare hillsides. The impact of falling rain, with a mean annual erosivity of 10-15 kJ m^{-2} (Morgan, 1979) soon created a badlands topography.

At Serdang, 30 km from Kuala Lumpur, between 1170 and 1500 m^3 km^2 y^{-1} of soil were lost from experimental bare soil plots (Morgan *et al.*, 1982). The estimated total sediment yield to the Sungai Anak Ayer Batu during the land clearance and construction stage in 1969-70 was 800 m^3 km^2 y^{-1} (Douglas, 1978). Depth-integrated suspended sediment samples collected by Leigh (1982) in the Sungai Anak Ayer Batu, 2.5 km below the site for which the 1969-70 estimate was made, show that the sediment problem persisted. In 1969-70 the range was 55-81230 mg 1^{-1}, in 1977, 10.4-15343 mg 1^{-1}. Much came from small intensely eroded areas; a 6.3

hectare deeply gullied, bare soil area yielded some 230 000 m^3 km^2 y^{-1} of sediment between 1970 and 1977 (Leigh, 1982).

These great volumes of sediment cause channel adjustments. When flood peaks and sediment loads increase sufficiently, river channels undergo a complete change of morphology (river metamorphosis) (Schumm, 1977). Thus the Sungai Anak Ayer Batu, which in 1969 was widening its channel and accumulating sediment in its bed, changed from a meandering, relatively deep and narrow stream with a gentle gradient, to a braided, wide, shallow and relatively straight, steep channel. The channel tended to enlarge and cut off bends, causing severe bank erosion downstream of the site. Flooding occurred frequently downstream on the University of Malaya campus and in the adjacent Pantai Valley. Its sudden pulses of sediment temporarily raised the bed of the main Sungai Kelang reducing channel capacity and causing local flooding (Douglas, 1975). Downstream, landholders had to protect their property.

As the housing estates became established in the late 1970s, some of the tributary channels were converted into concrete-lined storm drains according to the government-approved design criteria.

Once the land surface was covered with houses and gardens, sediment yields decreased but, with a large proportion of the catchment now impervious, run-off is rapid. The enlarged stream channel downstream of the storm drains is now maintained by flood flows. Water discharging from the drains has sufficient energy to pick up more sediment, unless its velocity is drastically reduced by increased channel roughness. Thus at present, there is scour immediately below any reach where banks have been lined with concrete or stones. The channel remains unstable as it adjusts to the rapid urban run-off. Storm water drainage works upstream have not solved the problem for downstream landholders.

Urban growth and rivers on great plains

Problems of a different magnitude exist for towns and cities in the plains traversed by great rivers. In the anabranching river systems of the inland deltas of the Niger, Nile and Darling River systems distributary channels may overflow in rare floods, creating temporary broad expanses of surface water across the imperceptibly sloping alluvial plains. Towns in such areas can be flooded to depths of over 1 m. Local protective embankments may merely

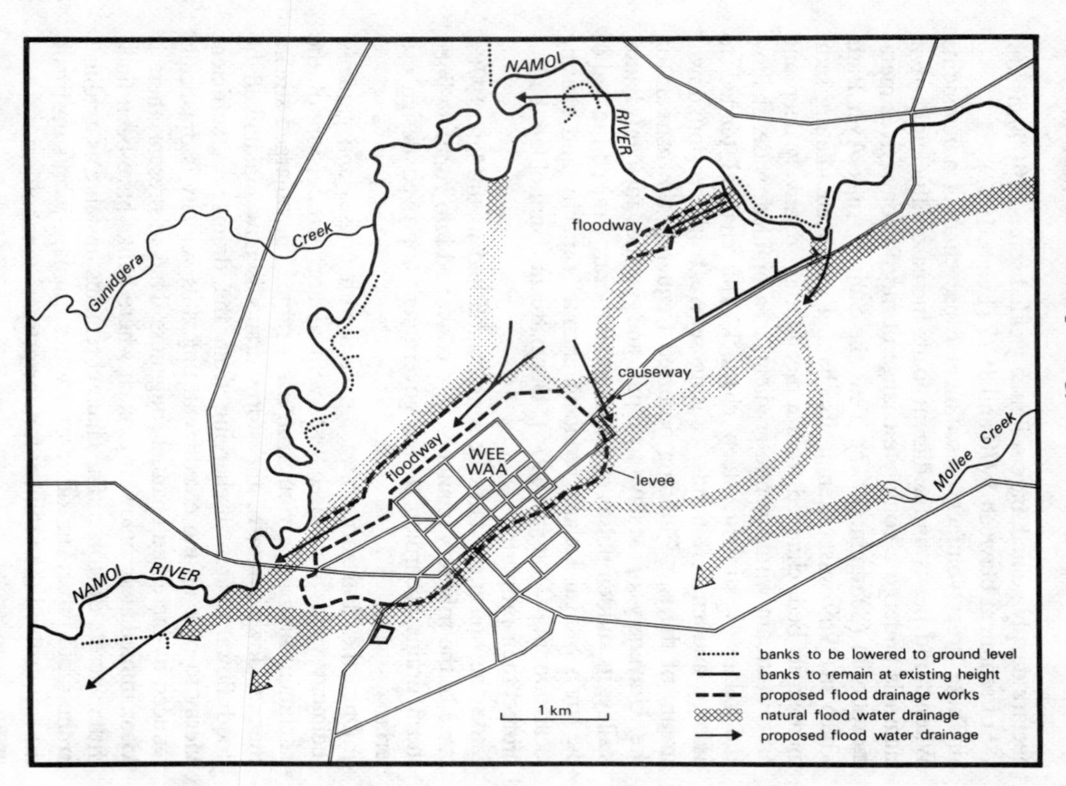

Figure 14.3: Proposed floodways and flood drainage works around Wee Waa, New South Wales. (Based on data from the Water Resources Commission, New South Wales.)

divert water to seldom flooded areas, causing severe disruption and damage (Douglas and Hobbs, 1979). Air photograph interpretation helps river managers to identify the distributary channels and natural floodways, and to plan levée banks to protect towns like Wee Waa (Figure 14.3) in the Namoi Valley cotton-growing area of the Darling Basin.

Cities and glacial deposits

Many cities are built on varied, unconsolidated sandy, gravelly or clayey deposits which were laid down as the last ice-sheets retreated, or which resulted from the periglacial reworking of earth surface materials. Foundation problems stem from the physical properties of the materials themselves, and from their vertical and lateral variability. For example, at the M56/M63 motorway interchange in the Mersey Valley, boreholes revealed two tills, buried river channels and other lithological complications. Distinguishing the glacial sands and gravels from the underlying Bunter sandstones was not easy, but careful investigations meant that only 18 per cent of the piles required more work than originally envisaged (Tate, 1976).

A more difficult engineering problem has been posed by glacial deposits where the River Irwell meanders against The Cliff, a cut in a fluvioglacial ridge between Lower Kersal and Higher Broughton, north of Manchester. The undercutting of a sequence of sands and clays has caused mass movements since 1882, when Great Clowes Street was first damaged. Groundwater moving through the sands eluviates the finer material, creating voids which lead to subsidence of the overlying material. Parts of the undercut slope also slump towards the river. In 1925 trams were diverted from Great Clowes Street between Hope Street and Bury New Road to reduce heavy traffic, and sections of streets adjacent to the cliff edge were closed. None the less, between 1931 and 1955, subsidence of 30-40 cm occurred in several places between the top of the cliff and Bury new road to the east (Figure 14.4). Further streets were closed in 1966, and in 1976 and 1977 the areas subsided further. Salford City Council has now exercised strict planning controls in The Cliff conservation area, a distinctive historic residential area.

Careful mapping of glacial deposits on the site of Washington New Town in Tyne and Wear also guided the selection of suitable

Figure 14.4: The Cliff, Higher Broughton, Greater Manchester, showing areas affected by slope instability and subsidence.

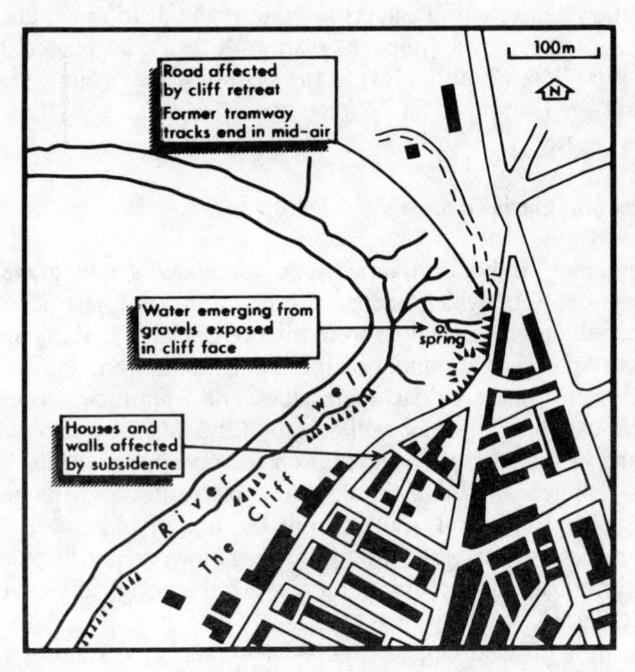

locations for the various new town land uses (Woodland, 1968). Such mapping would have included the site for the much publicised new Nissan car plant.

Similar problems were encountered on Beacon Hill, Boston, USA. The hill was mistaken for a drumlin, but as the digging for an underground car park proceeded, three gravel anticlines were revealed. Costly drainage works were necessary to cope with the large flow of groundwater through the gravels (Kaye, 1976).

Cities and subsidence

The term 'subsidence' describes human-induced lowering of the ground surface, due to abstraction of groundwater, oil and gas, removal of salt, coal and other minerals, and pumping of water from mines. Also, roofing and paving the land surface result in less replenishment of groundwater by infiltration, further lowering water-table levels.

Figure 14.5: Tramlines ending in mid-air where collapse of Great Clowes Street at The Cliff, Higher Broughton, has been caused by mass movement and subsidence in glacial sands undercut by the River Irwell.

One of the effects of removing coal or other minerals from a seam is to create a basin-like depression at the surface unless measures to prevent subsidence are taken. The maximum settlement is never as great as the thickness of the seam, although the surface area affected may be greater than the area of the underground workings. As buildings come within the limits of the subsiding basin, tension stresses tend to increase their length; closer to the centre, compressional forces tend to buckle and shorten them. Damage from mining subsidence is therefore expressed in terms of changes of length of a structure (Table 14.3).

The degree of damage to the surface is governed by three factors: orthodox ground movements related to the dimensions of mineral extraction; geotechnical conditions, such as the stability of the site prior to subsidence; and the extent to which buildings and other structures on the surface can withstand ground movements (Shadbolt and Mabe, 1970). Careful consideration of the total environmental impact of subsidence mitigation methods is necessary. In Britain, provision for compensation is made under an Act of Parliament, but the National Coal Board or other agency involved has to prove that the subsidence is due to mining before claims are met or repairs carried out.

Table 14.3: National Coal Board Classification of subsidence damage

Change of length of structure (m)	Class of damage	Description of typical damage
< 0.03	Very slight or negligible	Hair cracks in plaster. Perhaps isolated slight fracture in the building, not visible on outside
0.03-0.06	Slight	Several slight fractures showing inside the building. Doors and windows may stick slightly. Repairs to decoration probably necessary
0.06-0.12	Appreciable	Slight fracture showing on outside of building (or one main fracture). Doors and windows sticking: service pipes may fracture
0.12-0.18	Severe	Service pipes disrupted. Open fractures requiring rebonding and allowing weather into the structure. Window and door frames distorted; floors sloping noticeably; walls leaning or bulging noticeably. Some loss of bearing in beams. If compressive damage, overlapping of roof joints and lifting of brickware with open horizontal fractures
>0.18	Very severe	As above, but worse, and requiring partial or complete rebuilding. Roof and floor beams lose bearing and need shoring up. Windows broken with distortion. Severe slopes on floors. If compressive damage, severe buckling and bulging of the roof and walls.

Some of the most serious subsidence problems are occurring in the burgeoning cities of the Tropics. At Mexico City subsidence first caused concern in the 1850s but, by 1939, total subsidence nowhere exceeded 1.5 m. However, after 1939, groundwater abstraction accelerated leading to increased subsidence of up to 20 cm per year (Fox, 1965). The Carlos IV Monument (point C, Figure 14.6) is now more than 7 m below the original surface level. Subsidence is uneven in space and time, the Alameda gardens (point B) sinking less than the surrounding built-up areas. The area north-east of the cathedral (point A) has subsided less because it coincides with the site of the main square of the Aztec city of Tenochtitlan where the rubble of the destroyed buildings formed material and subsoil of the post-Hispanic city.

The subsidence has greatly increased the flood hazard in Mexico City. Removal of stormwater run-off has required massive urban drainage works. Mexico City lies in the internal drainage basin of Lake Texcoco and excess water has had to be diverted by

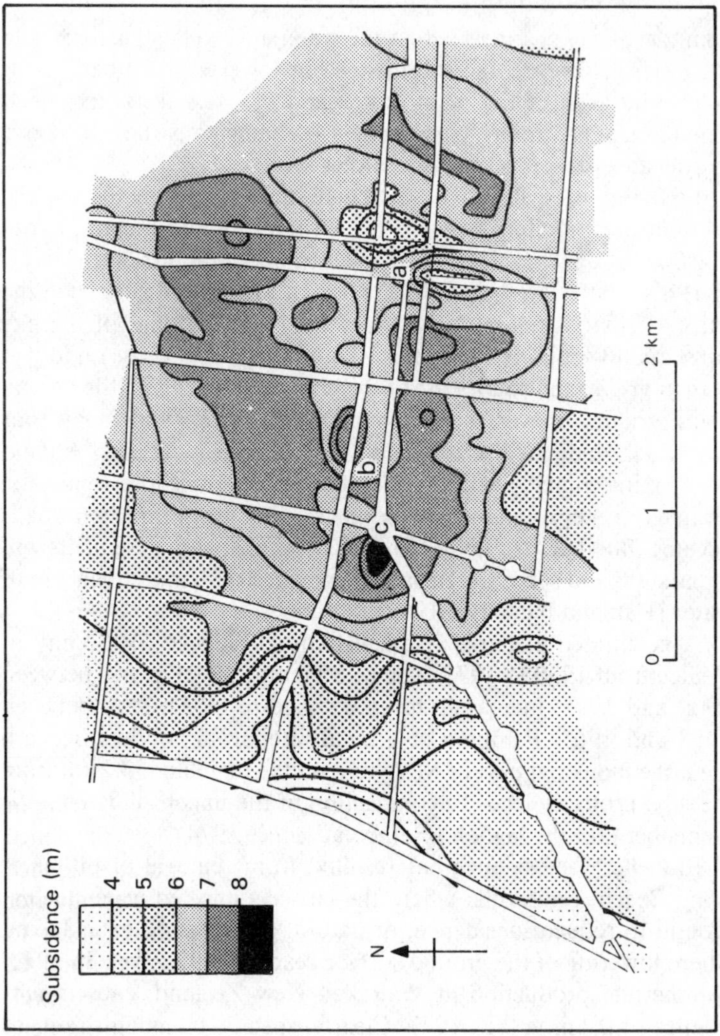

Figure 14.6: Isopleths of subsidence from 1891 to 1966 in central Mexico City; (a) Cathedral; (b) Alameda Central; (c) Monumento a Carlos IV (based on Hirart and Marsal, 1969).

taking artificial channels across the watershed to the Tula River to the north. However, the subsidence has lowered the urban drains below the level of the main drains and has also disrupted levels along individual drains. New tunnels are being constructed to carry stormwater run-off 40 m below street level.

In Bangkok, wells have also been sunk to exploit the alluvial groundwater aquifers beneath the city. Water abstraction, about 700 000 m^3 day^{-1} in 1974, had risen to about 1 000 000 m^3 day^{-1} in 1982. Water levels were within 1 m of the ground surface in the early 1950s but fell to 40 m below by 1978; 285 km^2 of the city consequently subsided irregularly some 0.8 m, but most rapidly at 0.14 m yr^{-1} to the south-east of the city. Faults beneath the coastal plain produce uneven movements. Drains, sewers and canals thus have their gradients disrupted and their efficiency reduced. Should heavy rains coincide with high tides and strong onshore winds, the drainage system cannot handle the urban run-off. Main roads became flooded; traffic is paralysed; shops and basements are inundated; and some streets may be covered with more than 1 m of water (Rau and Nutalaya, 1982).

A celebrated example is the 0.15 m subsidence of the city of Venice from 1930 to 1973. The rate averaged 0.01 m yr^{-1} between 1909 and 1925, increasing to an average of 0.05 m yr^{-1} between 1953 and 1961. Pumping of groundwater from aquifers beneath the large industrial port of Marghera on the mainland, 3.2 km from the city, probably leads to compaction of the unconsolidated sediments beneath the lagoon (Gambolati *et al.*, 1974).

Elsewhere, subsidence has resulted from removal of oil, as at Long Beach, California, where the city has injected water underground to reduce subsidence; or natural gas, as at Niigata, Japan, where lowering of the ground surface reached 0.32 m yr^{-1} in 1960. Geothermal production at Wairakei, New Zealand, caused subsidence of 0.1 m yr^{-1} in 1977. Clearly, specific planning strategies are essential in subsidence-prone areas.

Cities and karst

In karst, solution along bedding planes, fractures, joints and lithologic contacts creates preferential zones of water movement (Jennings, 1971). Karst water regimes, therefore, pose special urban problems, related both to the structural stability of the bedrock and to contamination of groundwater (Sheedy *et al.*, 1982). Structural failure may result from intensive pumping of groundwater which withdraws support in solution cavities, or from concentration of infiltration of surface water at the edges of paved areas or at roof gutter outlets and from other changes in water flows. Increased weight on the ground surface by landfill or building construction is another factor.

Natural sinkholes in karst pose particular problems. In southeastern England, collapses in the unconsolidated sediments overlying vertical pipes in the chalk occur periodically as a result of fluctuations in groundwater or changes in surface loading (West and Dumbleton, 1972). Such collapses, often affecting the overlying Tertiary strata, perhaps developed through solution by seepage water collected in sand and gravel lenses in the Reading beds above the chalk. These features are still active or potentially active hazards, but often with no surface topographic expression, as encountered in the construction of the A3(M) highway across the escarpment of the South Downs.

In the humid Tropics, karst tends to have more depressions than upstanding surface. Thus, in peninsular Malaysia, alluvium on karst margin plains (Jennings, 1971) mantles a complex of karst pinnacles with a relief of 10 m or more. Boreholes only 1 or 2 m apart may hit different pinnacles giving the impression of a continuous limestone surface, whereas in reality the two limestone summits may be separated by a depression over 10 m deep. When the new bridge over the Kelang River was being constructed for the Federal Highway linking Kuala Lumpur with the airport and with Port Kelang, the major foundation problem was that of the buried pinnacled karst surface.

Other problems in karst plain urbanisation include increases in the frequency and magnitude of sinkhole flooding, and the probability of collapse. In Florida, soil washed off housing sites accumulates in the depressions, reducing seepage through the sinkhole and inducing sinkhole-related flooding. Kemmerly (1981) has suggested that federal mortgage agencies should recognise the flood hazards associated with sinkholes so that losses from this type of flooding may be minimised.

Most urban areas rely on sanitary landfill for solid waste disposal. However, in karst terrains infiltrating water penetrating such a dump may leach out noxious substances which then move rapidly through solutional passages. Rapid contamination of the karst water system can ensue, affecting water used for domestic and industrial purposes. Leakages from septic tanks act in a similar way to landfill leachates. Vogt (1972) showed that seepage led to an infectious hepatitis outbreak at Posen, Michigan. In Florida, hydrologic maps have been used to delimit relatively favourable sites for sanitary landfills. Such sites are found in areas where the top of the Floridan aquifer is more than 8 m below the land surface and is overlain by relatively impermeable beds; having adequate surface drainage not subject to flooding; outside the drawdown area of large groundwater pumping systems; remote from streams connected with the Floridan aquifer; and with no history of sink-hole formation.

Cities and beaches

Coastal zones, especially beaches, are both attractive and hazardous. Many coastlines are made up of dune ridges the stability of which is ensured until the dune vegetation or the dynamics of wave and sand movement are altered. On the coast of Belgium and the Netherlands, which is similar to that of South Lancashire, expansion of towns like Oostende and Zeebrugge has affected the dune system; those between Zeebrugge and Knokke have migrated inland, medieval village sites now being several hundred metres out to sea. Major storm surges occasionally breached the dunes, and by the late sixteenth century groynes had been constructed at Heist. Deepening of the Appelzak trench offshore led to increased beach erosion. Later, construction of sea-walls, roads and buildings on the dunes broke the dynamic equilibrium between the dunes and the beach by restricting the supply of sand from the dunes to the beach. Thus, instead of receding with a similar profile, the beach has become steeper, and erosion has been accelerated. Although the sea-wall prevents storm waves attacking the dunes, the energy of wave impact is spent on undermining the beach at the base of the wall. Completion of the Zeebrugge mole disrupted the replenishment of sand by longshore currents. The mole trapped the sand carried by the floodtide on its western side, while that removed by the ebbtide silted up the harbour to the east, and the natural beach at Heist and Knokke virtually disappeared. Replenishment by pumping sand from dunes inland was necessary to retain the fundamental, basic attraction of a seaside resort.

Similar problems arose at Santa Monica, California, where the residents wanted a sheltered harbour. To avoid interrupting the eastward movement of sand along the coast, a breakwater was constructed parallel to the shore in 1933. However, a 'wave shadow' area developed, with the beach behind the breakwater growing outwards and trapping the eastward-moving sand. Eventually, erosion of beaches further east occurred with the undermining of houses built on the foreshore. Individual efforts to protect homes led to further beach erosion east of each site so treated. None the less, urban development on some of the barrier islands of the east coast of the United States has been rapid, increasing from 5.5 per cent of the barrier island area in 1945 to 14 per cent in 1974 (Lins, 1981). At Ocean City, Maryland, most of the sand seaward of the present line of development has been eroded away. Beaches can now only be maintained by artificial replenishment. Annually, as the beach migrates inland, each building in turn becomes exposed to the waves at the beach front.

Cities and expansive soils

Excessive shrink–swell behaviour can produce cracking and failure of building walls, foundations and floors. The main clay mineral involved is montmorillonite. The effects of such soils are particularly marked in semi-arid and sub-tropical areas, but are also fairly common in Southern Britain. Therefore, the Building Research Station buried a series of plates 0.3 m, 0.6 m, 0.9 m and 1.2 m below the surface of a boulder clay at their research station near Watford. At 0.3 m below the surface 30 mm of vertical movement was recorded, but as little as 5 mm at a depth of 1.2 m. Slight year-to-year differences were noted, a wet autumn producing 25 mm of vertical movement, a drier autumn only 20 mm.

The exceptionally dry years of the 1975-76 British drought saw many cases of foundation movements resulting from clay sinkage. A spate of insurance claims followed (Table 14.4), but they probably over-exaggerate the damage, as changes in insurance regulations affected people's attitudes to making claims.

Among the causes of the cracking in Table 14.4 are, of those associated with the ground, ground subsidence and heave due to volume changes in clay soils; settlement and heave of floor slabs or unsuitable or poorly compacted infill beneath the slab; instability

Table 14.4: Annual number of claims for cracking of buildings notified to 15 major insurance offices in Britain 1971-79

1971	164
1972	509
1973	817
1974	1705
1975	3821
1976	20922
1977	112511
1978	7795
1979	7454

Source: *Building Research Establishment Digest, 251.*

of sloping ground; movement due to consolidation of poor ground or made ground; mining subsidence movement caused by nearby excavations; and chemical attack on foundation concrete.

In lower latitude cities the problems are more severe. Householders in Dallas County, Texas, experience approximately 8470 failures of property foundations annually. Of these, 84 per cent occur in areas with soils containing a high percentage of montmorillonitic clays. The total annual cost of failures on such soils has been estimated at almost $12 million out of a total of $14 million for all foundation failures in Dallas County (Petak and Atkisson, 1982). Expansive soils are also a significant, widespread geomorphic hazard in the Front Range urban corridor around Denver, Colorado (Hansen and Crosby, 1982). In 1974, the annual cost of damage from expansion soils was estimated at $2.3 billion (Gromko, 1974).

Study of the soil characteristics of any urban site is essential in construction planning on expansive soils. Mitigation of potential damage may be achieved by chemical means, presaturation, compaction, and the design of special pier, grade-beam or reinforced slab foundations. Also, despite the many benefits they bring to cities, rapidly growing trees accentuate shrinkage, and can abstract soil water from beneath the paving.

Conclusion

Illustrations of the geomorphic problems of cities demonstrate that the appropriate investigations, remedial measures, or land use decisions can minimise difficulties. However, since urban development is a type of geomorphic change, land capability assessment techniques incorporating geomorphic and other earth science information are desirable. In several cities, local bye-laws ensure that some geomorphic hazards are avoided. Such legislation restricts building on steep slopes or requires that developers do not allow the run-off or sediment yield from construction to exceed those which existed before construction began. However, extension of such legislation requires sufficient awareness of the significance of the problem. An enlarging theme of geomorphology, therefore, is the education of the public, lawyers and politicians about the problems, and the ways in which they can be addressed.

References

Coates, D.R. (1983), 'Large-scale land subsidence', in Gardner, R. and Scoging, H. (eds), *Mega-geomorphology* (Clarendon Press, Oxford), pp. 212-34.

Douglas, I. (1975), 'The impact of urbanisation on river systems', *N. Z. Geog. Soc. Conf. Ser., 8*, 307-17.

Douglas, I. (1978), 'The impact of urbanisation on fluvial geomorphology in the humid tropics', *Geo-Eco-Trop, 2*, 229-42.

Douglas, I. and Hobbs, J.E. (1979), 'Public and private reaction to three flood events in north-western New South Wales, 1971-1976', in Heathcote, R.L. and Thom, B.G. (eds), *Natural Hazards in Australia* (Australian Academy of Science, Canberra), pp. 251-9.

Fox, D.J. (1965), 'Man–water relationships in metropolitan Mexico', *Geographical Rev., 55*, 523-45.

Gambolati, G., Gotto, P. and Freeze, R.A. (1974), 'Predictive simulation of the subsidence of Venice', *Science, 183*, 849-51.

Gromko, G.J. (1974) 'Review of expansive soils', *J. Geotech. Eng. Div. Am. Soc. Civil Engineers, 100(GT6)*, 667-87.

Hansen, W.R. and Crosby, E.J. (1982), 'Environmental geology of the Front Range urban corridor and vicinity, Colorado', *US Geol. Survey Prof. Pap., 1230*.

Hawkins, A.B. and Privett, K.D. (1981), 'A building site on cambered ground at Radstock, Avon', *Q. J. Eng. Geol., 14*, 151-67.

Hiriart, F. and Marsal, R.J. (1969), 'The subsidence of Mexico City', in Carillo, N. (ed.), *The Subsidence of Mexico City and Texcoco* (Secretaria de Hacienda y Credito Publico, Mexico), pp. 109-47.

Jennings, J.N. (1971), *Karst* (ANU Press, Canberra).

Kaye, C.A. (1976), 'Beacon Hill end moraine, Boston: new explanation of an important feature', in Coates, D.R. (ed.), *Urban Geomorphology, Geol. Soc. Am. Sp. Pap., 174*, 7-20.

Kemmerly, P. (1981), 'The need for recognition and implementation of a sinkhole–floodplain hazard designation in urban karst terrains', *Environ. Geol., 3*, 281-92.

Leigh, C.H. (1982), 'Urban development and soil erosion in Kuala Lumpur, Malaysia', *J. Environ. Management, 15*, 35-45.

Lins, H. (1981), 'Patterns and trends of land use and land cover on Atlantic and

Gulf Coast barrier islands', *US Geol. Survey Prof. Pap., 1156.*
Morgan R.P.C. (1979), *Soil Erosion* (Longmans, London).
Morgan, R.P.C., Hatch, T. and Wan Sulaiman Wan Harun (1982), 'A simple procedure for assessing soil erosion risk: a case study for Malaysia', *Z. Geomorph. Suppl., 44,* 69-88.
Mosley, M.P. (1975), 'Meander cutoffs on the River Bollin, Cheshire, in July 1973', *Rév Geomorphol. Dynam., 24,* 21-31.
Petak, W.J. and Atkisson, A.A. (1982), *Natural Hazard Risk Assessment and Public Policy: Anticipating the Unexpected* (Springer-Verlag, New York).
Rau, J.L. and Nutalaya, P. (1982), 'Geomorphology and land subsidence in Bangkok, Thailand', in Craig, R.G. and Craft, J.L. (eds), *Applied Geomorphology* (George Allen & Unwin, London), pp. 181-201.
Richards, K.S. (1982), *Rivers: Form and Process in Alluvial Channels* (Methuen, London).
Schumm, S.A. (1977), *The Fluvial System* (Wiley-Interscience, New York).
Shadbolt, C.H. and Mabe, W.J. (1970), 'Subsidence aspects of mining development in some northern coalfields', in *Geological Aspects of Development and Planning in Northern England* (Yorkshire Geological Society, Leeds), pp.108-23.
Sheedy, K.A., Leis, W.M., Thomas, A. and Beers, W.F. (1982), 'Land use in carbonate terrain: problems and case study solutions', in Craig, R.G. and Craft, J.L. (eds), *Applied Geomorphology* (George Allen & Unwin, London), pp.202-13.
Tate, A.P.K. (1976), 'Pile driving in glacial deposits', *Q.J. Eng. Geology, 9,* 280.
Vogt, J.E. (1972), 'Infectious hepatitis outbreak in Posen, Michigan', in Pettyjohn, W.A. (ed.), *Water Quality in a Stressed Environment* (Burgess, Minneapolis), pp.188-93.
West, G. and Dumbleton, M.J. (1972), 'Some observations on swallow holes and mines in the chalk', *Quart. J. Eng. Geol., 5,* 171-8.
Woodland, A.W. (1968), 'Field geology and the civil engineer', *Proc. Yorks. Geol. Soc., 36,* 531-78.

15 RELIEF FORMS ON PLANETS

V.R. Baker

Introduction

In their review *Geographic Geomorphology in the Eighties*, Graf *et al.* (1980, p. 283) observed, 'Research on the surface forms and processes of other planets also provides an opportunity for geographers to work in a significant interdisciplinary environment, but so far geographers have been noticeable by their absence in this area'. Some would find this completely appropriate. Geomorphology, they would argue, is properly restricted either to the dynamic geology or to the physical geography of the surface of the earth. Indeed, more argument would probably be expended on the relative 'geologic' or 'geographic' content of geomorphology than on any consideration of bizarre, alien landscapes.

In the United States several prominent departments of Geology and Geophysics have eliminated the traditional faculty position of 'geomorphologist'. Instead, these departments have expanded programmes in what they perceive to be the most exciting fields of the earth sciences, including, for example, geophysics, tectonics, geochemistry, tectonophysics and oceanography. Of course, this trend is shortsighted. The importance of the planetary surface that we inhabit cannot be denied through arbitrary academic fads. Nevertheless, this disturbing trend raises a question: have geomorphologists properly conveyed to their colleagues the full excitement, the creativity, or the wonder of their science?

A century ago such questions never arose. Geomorphology was a science filled with wonder and excitement. The stimulus for its initially rapid growth was the discovery of landscapes which seemed then as bizarre and alien as those on other planets. The great plateaux of the Western United States, the hyper-arid deserts of the Eastern Sahara, the steppes of Central Asia, the karst of Dalmatia, and the inselbergs of Australia and Africa all posed anomalies in the prevailing geomorphic theory. The explanation of the new landscapes led to an expanded and improved explanation for landscapes already described.

Bloom (1978, p. xvi) predicted 'that the schemes of explanatory description that are ultimately used for submarine and planetary landscapes will not be very different from the one developed for describing subaerial terrestrial scenery'. This reflects a fundamental quality of science: there are no bounds, geographic or otherwise, for inquiry into the origin of phenomena. If a geomorphologist can learn more about the surface of the earth by studying other planetary surfaces, then that extraterrestrial study can no longer be dismissed as an interesting intellectual diversion: it becomes an essential part of geomorphology. An earth-centred view of geomorphology is as limiting as a pre-Copernican view of the solar system.

Not all geomorphologists have shared the modern reluctance to consider the study of extraterrestrial relief forms. In 1892, the US Geological Survey suffered a drastic cut in its research funds. Their chief geologist at that time was the geomorphologist Grove Karl Gilbert. Without support for his field work, Gilbert undertook a study using the US Naval Observatory telescope in Washington, DC, to compare surface features on the moon with counterparts on earth (El-Baz, 1980). Despite the prevailing view that gradual and prolonged volcanism explained lunar surface features, Gilbert (1893) concluded that cataclysmic impact processes best explained the ubiquitous lunar craters. He applied the term 'meteoric' to his impact theory, which had to wait over 70 years for verification by the Apollo programme of lunar landings and sample returns. Among the many lessons that geomorphology can trace to Gilbert's example (Baker and Pyne, 1978) that of studying other planets besides the earth has yet to be fully appreciated.

Scales of study

Cailleux and Romani (1981) believe that there are two major trends in modern geomorphology: towards qualification and towards more varied extensions to other scientific disciplines, to applied problems, to longer time-scales and more ancient features. They also see an extension to greater spatial scales. The study of planetary landforms can be considered a part of this trend. As pointed out by Sharp (1980, p. 231), 'One of the lessons from space is to "think big"'. Planetary landforms are immense features indeed (Table 15.1). Moreover, the processes responsible for them

Table 15.1: Spatial scales of some planetary landforms

	Earth	Moon	Mars
Impact craters (basins)	200 km	2400 km	2000 km
Landslides	10 × 20 km	2 × 10 km	100 × 200 km
Thermokarst	10 km	—	100 km
Fluvial channels (flood)	10 × 200 km	—	100 × 2000 km
Erg (sand sea)	300 × 1000 km	—	200 × 5000 km
Shield volcano	120 km wide 9 km high	25 km wide 2 km high	700 km wide 25 km high

exhibit immense expenditures of energy (Table 15.2). These scales contrast with modest processes, whose study has dominated modern geomorphic concern. When most geomorphologists abandoned the Davisian model, and its attendant concern with denudation chronology, many embraced a systems approach, with emphasis on statistical analysis and predictive modelling of process–response phenomena on the earth's surface. However, this change in methodology also involved a change in the scale of phenomena studied. Small-scale features and short-acting processes proved most amenable to the new methodology. The global concern with long-acting denudation was ignored, not because of its importance, but because of methodological inadequacies in its past study.

The concept of scale is so fundamental to geomorphic methodology that it is the focus of a current philosophical debate. On the one hand, Thornes and Brunsden (1977, p. 116) state, 'The current paradigm is one in which process studies prevail effected principally and increasingly through mathematical and stochastic models'. On the other hand, Church (1980) writes, 'Contemporary

Table 15.2: Scales for some planetary geomorphic processes

Process	Time (s)	Velocity (cm/s)	Power (ergs/cm^2-s)
Meteor impact (1 km)	10^{-1}	10^6	10^{15}
Sturzstrom	10-10^2	5×10^3	10^6
Missoula flood	10^4-10^6	10^3	10^7
Wind storm	10^4-10^6	10^4	10^4
Amazon River	10^4-10^7	10^2	10^4
Glacier	10^7-10^{10}	10^{-5}	1

process studies are of little worth in evaluating landscape evolution'. The fact remains that landscape evolution remains a critical concern for geomorphology despite past problems in the implementation of its study. Thornbury (1969) made a key point when he noted that geomorphology, although concerned primarily with present-day landscapes, attains its maximum usefulness by historical extension. In the phenomenal world of interplanetary landscape comparison, that historical extension is measured in billions of years. Geomorphic data on ancient relict planetary surfaces are critical in interpreting the early histories and evolution of planetary surfaces, atmospheres, hydrospheres and biospheres throughout the solar system.

Methodological themes

Arguably, the methodologies of a science are reflected in the research techniques of its practitioners. Indeed, the recent review by Goudie (1981) clearly demonstrates the methodological focus of geomorphology on small-scale, short-duration process studies.

Techniques

Most planetary geomorphology is accomplished by the interpretation of remote-sensing imagery. A variety of systems are available (Saunders and Mutch, 1980), including vidicon imaging, multispectral scanning, radiometers and radars. Modern image processing of digitally formated data has revolutionised the interpretation of large-scale planetary landscape scenes. Many geomorphologists have been slow to appreciate the global perspective afforded their own planet by these advances, let alone the study of other planetary surfaces. In any science new techniques are not important of themselves. It is rather the new discoveries made possible because of those techniques that stimulate scientific progress. A profound example of such a new discovery in terrestrial geomorphology came in November 1981 when the shuttle Columbia trained a space-age instrument on the earth. The shuttle-imaging radar (SIR-A) carried by Columbia produced radar images of the hyper-arid Selima sand sheet of the Eastern Sahara (Elachi *et al.*, 1982). The radar penetrated the sand cover to reveal fluvial valleys now filled by aeolian sand. The valleys discovered by radar interpretation show a regional drainage system formed when the modern aeolian-dominated landscape was subject to extensive fluvial erosion, probably during pluvial episodes of the Pleistocene and Tertiary (McCauley *et al.*, 1982).

Analogic studies

The study of planetary surfaces relies heavily on analogic reasoning to reconstruct the complex interactions of processes responsible for the observed landforms (Mutch, 1979). Thus, the photointerpreter of planetary images must rely on his experience with terrestrial landscapes. Moreover, the geomorphic interpretation of other planets produces a kind of intellectual feedback: some planetary surfaces contain excellent analogues for little-understood terrestrial processes. Sharp (1980, p. 231) emphasises this benefit as follows:

> Planetary exploration has proved to be a two-way street. It not only created interest in earth-surface processes and features as analogues, it also caused terrestrial geologists to look on earth for features and relationships better displayed on other planetary surfaces.

Among his many other remarkable studies, Gilbert also made one of the earliest studies of a terrestrial analogue to an extra-terrestrial landform. However, the study is exemplary for its method rather than its conclusion. He studied Meteor Crater in Northern Arizona (Figure 15.1), which was then called Coon Butte. With impeccable logic and the available scientific tools of his day, Gilbert (1896) concluded that the feature resulted from a steam explosion. Nevertheless, the meteor impact origin of the feature was not established incontrovertibly until 60 years later when two high-pressure forms of silica were found in the shock melts formed when an object about 25 m in diameter struck with an energy equivalent to a 2 megaton thermonuclear blast (French, 1977).

Despite his rejection of the impact origin, Gilbert did correctly observe that his conclusion was based on insufficient data. His criticism of his own results contains an important lesson for both terrestrial and planetary geomorphologists (Gilbert, 1896, p. 12):

> This illustrates the tentative nature, not only of the hypotheses of Science, but of what Science calls results. . . . However grand,

however widely accepted, however useful its conclusion, none is so sure that it can not be called in question by a newly discovered fact. In the domain of the world's knowledge there is no infallibility.

Gedankenexperiment

If, like physicists or chemists, geomorphologists were able to do real experiments on their objects of study, they would first define the important parameters of their experimental system. Then they would systematically change each parameter, one at a time, while holding all other system parameters constant. Of course, this can actually be done in some artificial systems, for example flumes and wind tunnels. Such artificial systems may even function as useful

Figure 15.1: Meteor Crater in northern Arizona provides an excellent analogue to the impact features on other planets. This 1.2 km crater formed about 25 000 years ago when an iron mass struck at a speed of 11 km/s yielding an energy equivalent of between 5 and 20 megatons of TNT (Shoemaker, 1981, p. 35). (Photograph by V.R. Baker.)

models, even though no such simplification has ever provided a full satisfactory explanation for that magnificent prototype that we call the 'landscape'. Because of the complexity of the modern landscape, its explanation requires geomorphologists to resort to thought experiments, or what the German philosophers call *Gedankenexperiment.* Geomorphologists have expended considerable energy in their analysis of theory as an element of science. Scientific theory is an elusive concept, more a part of collective scientific experience than an entity to be described and classified. True theory binds diverse consequences together in such an elegant manner that it compels belief by the scientific community (Judson, 1980). Because bold prediction is a major part of theory development, scientific models that make testable predictions comprise an important element in the theoretical framework of a science. In planetary studies, model-building, prediction and testing proceed at an exhilarating pace. Prior to the encounter of the Voyager I spacecraft with Io, a satellite of Jupiter, Peale *et al.* (1979) predicted that the dissipation of tidal strain energy on that moon would provide sufficient heat for intense volcanic activity. The Voyager spacecraft analysed volcanic plumes on Io rising as high as 300 km and spreading pyroclastic deposits up to 600 km from the vents (Strom and Schneider, 1982).

Fundamental concepts

Most geomorphologists would agree that certain fundamental assumptions underlie all geomorphological investigations. Whether termed 'fundamental concepts' (Thornbury, 1969), 'philosophical assumptions' (Twidale, 1977), or 'basic postulates' (Pitty, 1982), these ideas constitute a 'conventional wisdom' for the science. One such fundamental concept involves the inherent complexity of landscapes, which has impeded the development of grand theories that survive the test of explaining numerous local features. Another basic assumption involves climatic morphogenesis, emphasising the role of climatically controlled processes of landform genesis. Of course, these assumptions also serve to focus controversy, as exemplified by Twidale's (1983) advocacy of structural control over climatic control of pedimentation, inselberg formation and planation. Even more controversial have been the assumptions of uniformitarianism and landscape evolution.

Cataclysmic geomorphology

Because many planetary surfaces have been relatively stable for billions of years, they preserve the effects of extremely rare, exceedingly violent processes. Such processes include impact cratering, sturzstroms (large avalanches of rock and debris) and cataclysmic flooding (Table 15.2). On earth, the evidence of such catastrophes is meagre because of rapid crustal recycling through plate tectonics and relatively high denudation rates. However, on the other terrestrial planets the results of these processes can be studied in great detail. This is of profound importance for earth studies. When cataclysmic processes have occurred on earth, their influence has been profound. The extinction of numerous organisms at the end of the Cretaceous because of a meteor impact is a case in point (Alvarez *et al.*, 1980).

Figure 15.2: Drainage map of the Martian volcano Hecates Tholus.

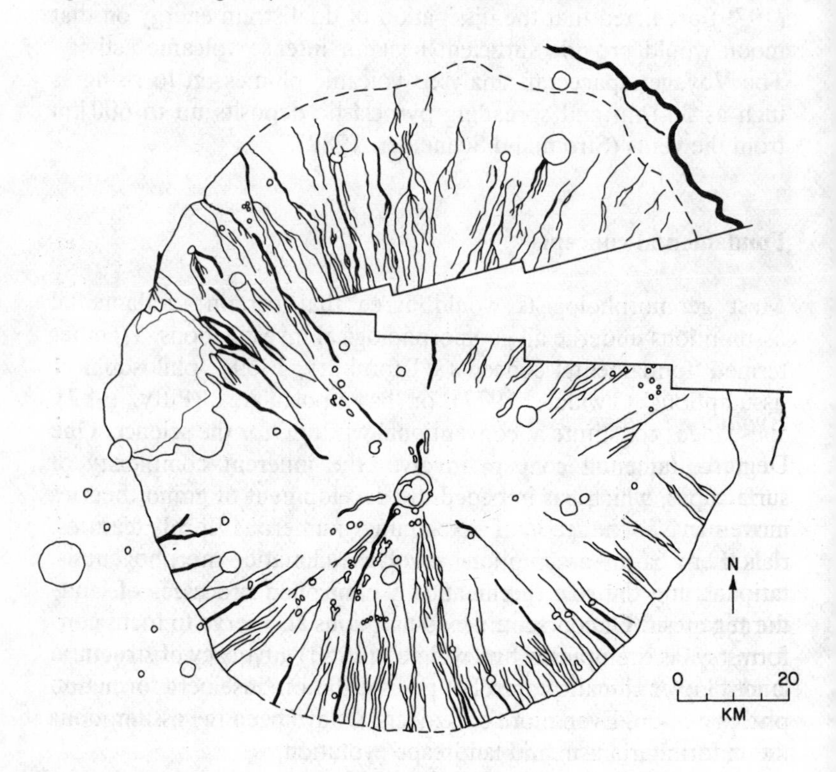

Figure 15.3: Viking orbital image of a structurally controlled valley network in the heavily cratered terrain of Mars. Note that the valleys are developed in an upland zone of heavily cratered terrain (HC) and indicate fluid flow towards a lowland zone of cratered plains (CP).

Evolution of landscapes

Degradation involves the wearing away or lowering of a planetary surface by natural processes of weathering, erosion and mass movement. It further involves the transportation of eroded materials away from eroding planetary terrains. The varying gravitational, atmospheric, thermal and geologic material parameters of the terrestrial planets have produced suites of degradational landforms reflecting varying modes of process operation. Planetary landscapes have experienced a time-directed change in morphology that depends on the intensity of process operation versus the resistance of the surficial materials. Thus, the classic Davisian trilogy of structure, process and time is an essential theme in planetary studies.

Structure. Planetary structures involve resistance to degradational processes. A volcano on Mars poses a primary structural form that is dissected by a radial pattern of valleys (Figure 15.2). The degradational pattern reflects the primary structure of volcanic flows on the shield volcano. Similarly, faults and joints may control valley development (Figure 15.3). Such patterns are well known on earth, but on Mars they are being interpreted to understand the geologic history of the planet. Indeed, Mars has so many analogues to earth landforms that it alone can be used to justify an interplanetary scope to geomorphology (Baker, 1981).

Process. Planetary landforms reflect the weathering, hillslope, fluvial, glacial, mass movement and aeolian processes responsible for their genesis. The list of planetary geomorphic processes is also the list of terrestrial geomorphic processes (Table 15.3). However, it is fascinating that some processes of limited interest on earth have proved to be of immense importance on other planets. Thermokarst, sapping and wind erosion are examples which have great relevance to Mars (Baker, 1981).

If river and hillslope processes indeed constitute the central theme of geomorphology (Leopold *et al.*, 1964, p. vii), then that theme surely applies to Mars and perhaps to other planets as well. The discovery of channels and valleys on Mars (Figure 15.4) poses exactly the same sort of fundamental concern for geomorphology

Table 15.3: Geomorphic processes on planetary surfaces

Process	Mars	Moon	Galilean satellites	Venus
Tectonic	R	R	R	R
Volcanic	R	R	A R	R
Weathering	A R	R	—	A R
Karst	Thermokarst	—	—	—
Hillslope	A R	R	—	R
Fluvial	R	—	—	?
Glacial (ice)	R	—	R	—
Aeolian	A R	—	—	A?
Coastal	R?	—	—	—

A — Active process; R — Relict process (landforms)

as did the question of valley origins considered by Hutton and Playfair in the formative age of the science (Baker, 1982). Fluvial geomorphology for Mars has only been in progress since 1971, but it has already produced a full exposition of scientific hypothesising and convergence on most probable explanations (Baker, in press). Water has clearly been demonstrated to be the primary agent of channel and valley formation on Mars (Baker, 1982; Mars Channel Working Group, 1983). Furthermore, this geomorphic fact is of importance to the possibility of life on Mars (Masursky *et al.*, 1979), constituting a profound extension of a geomorphic result to a fundamental question in biology.

Figure 15.4: Map of drainage networks in the heavily cratered terrains of Mars.

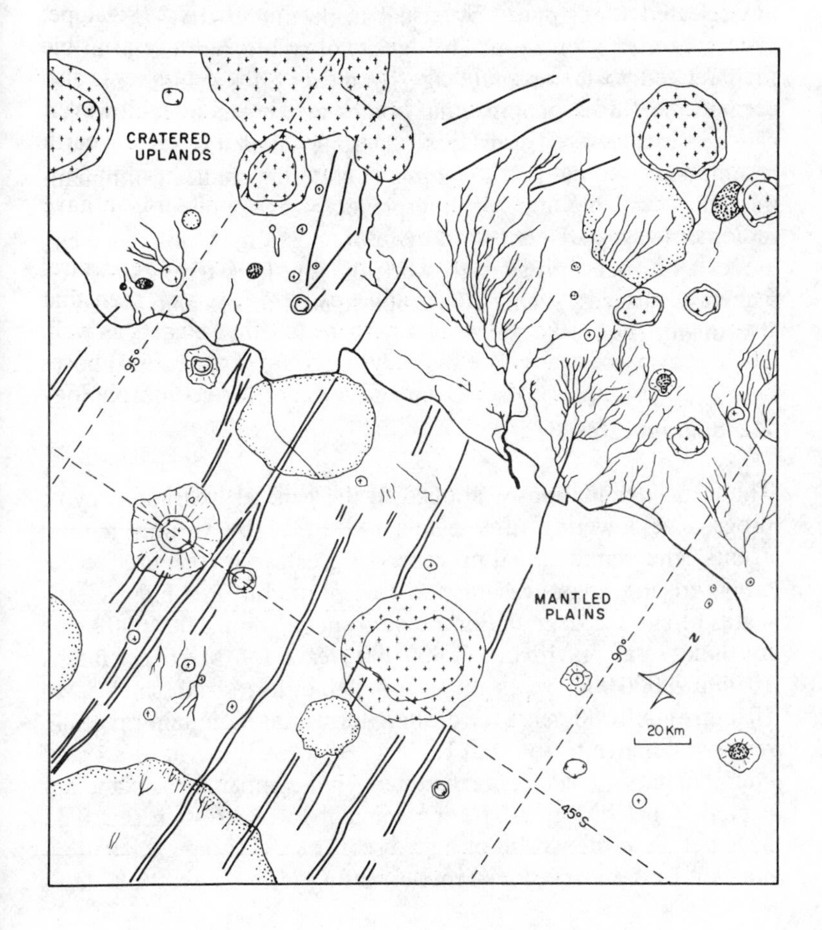

Time. The immense range of ages of extant planetary erosion surfaces (Table 15.4) and in rates of surface degradation on various planets allows the interplanetary comparison of fundamental controls on degradational processes through geological time. The degradational history of very ancient terrains on the terrestrial planets can even provide information on the early Precambrian history of earth. This unforeseen application of geomorphology parallels the recognition of extremely ancient landscapes and 'paleoforms' on earth (Twidale, 1976; Young, 1983). The use of new dating techniques for planetary surfaces is quantifying the neglected theme of denudation chronology. The use of cratering statistics to date planetary surfaces has evolved to a working methodology (McGill, 1977). Surface areas of a planetary body are selected that appear by visual inspection to have a homogeneous crater population. The density of craters on that surface is then related to an absolute age by means of assumptions concerning the fluxes of impacting bodies and the complex modifications of cratered surfaces. Despite limitations in these assumptions, the method of dating by cratering studies will remain the only available tool of interplanetary age comparison until radiometric ages are generated by future missions. Using this technique it has been possible to work out the long-term history of planetary surfaces, such as the emplacement of volcanic materials (Hartmann *et al.*, 1982).

Denudation rates

The term 'denudation' is applied to the general wearing away or progressive lowering of a planetary surface by various natural agents. The denudational processes of greatest interest are weathering, erosion, mass-wasting and transportation. Preliminary estimates of surface degradation range from <1 mm/1 000 000 years for lunar craters, 10^2 mm/1 000 000 years for Martian craters, 10^3 mm/1 000 000 years for very stable terrestrial shields, to 10^6 mm/1 000 000 years for eroding terrestrial mountain ranges in regions of active tectonism (Table 15.5). Clearly we have a grand interplanetary *Gedankenexperiment* in comparative planetary degradational histories. The impediment to progress in this field does not lie in the wealth of new discoveries awaiting explanation, nor is it in the marvellous remote-sensing technology available to

Table 15.4: Ages of some planetary geomorphic surfaces

Planet	Surface	Age (BY)	Dating method
Moon	Highlands	4.0-4.4	Radiometric
	Mare	3.0-3.9	Radiometric
Mars	Cratered uplands	4.0	Crater density
	Plains	3.0-3.9	Crater density
	Outflow channels	1.0-3.5	Crater density
	Olympus mons	< 0.5	Crater density
Earth	Ancient erosion	0.2-0.3	Radiometric
	Columbia plateau	10^{-2}	Radiometric
	Hawaii	10^{-3}	Radiometric
	Coastal plains	10^{-3}	Radiometric
	Riverine plains	10^{-5}	Radiometric

Table 15.5: Denudation rates on some planetary geomorphic surfaces

Planet	Process	Rate (mm/1 000 000 yr)	Reference
Moon	Horizontal transfer	8	Arvidson *et al.* (1979)
	Erosion of craters	1	Arvidson *et al.* (1979)
Mars	Pedestal craters	10^2	Carr (1981)
	Escarpment retreat	10^5	Carr (1981)
	Inselbergs, c. Australia	10^3	Twidale (1978)
	Valley incision, c. Australia	10^3	Young (1983)
Earth	Amazon Basin	5×10^4	Gibbs (1967)
	Himalayas	10^6	Ollier (1981)
	Marine cliff retreat	10^9	Ollier (1981)

study the relief forms on other planets. The only limitation to progress is a reluctance by many geomorphologists to appreciate how broad and exciting their science has really become.

Conclusions

The study of relief forms on other planets is a natural extension of the science of landforms on our own planet. Much of modern planetary geomorphology is merely a reawakening of interest in some classic geomorphic themes, including the evolutionary sequence of landscapes, the preservation of ancient landforms, the role of cataclysmic events, the variable rates of process application,

Anders Rapp is Professor of Physical Geography, University of Lund

Ian Reid is a Lecturer at Birkbeck College, University of London

Wilf Theakstone is Senior Lecturer in Geography, University of Manchester

Rob Warner is Senior Lecturer in Geography, University of Sydney

Brian Whalley is Senior Lecturer in Geography, The Queen's University of Belfast

Torao Yoshikawa is Emeritus Professor of Geography, University of Tokyo

INDEX